Fall Creek Books is an imprint of Cornell University Press dedicated to making available again classic books that document the history, culture, natural history, and folkways of New York State. Presented in new paperback editions that faithfully reproduce the contents of the original editions, Fall Creek Books titles will appeal to all readers interested in New York and the state's rich past. Some of the books published under this imprint reflect the sensibilities and attitudes of an earlier era; these views do not necessarily reflect those of Cornell University Press. For a complete listing of titles published under the Fall Creek Books imprint, please visit the Cornell University Press website: www.cornellpress.cornell.edu.

VANISHING IRONWORKS of the RAMAPOS

The Story of the Forges, Furnaces, and Mines of the New Jersey–New York Border Area

James M. Ransom

FALL CREEK BOOKS
AN IMPRINT OF CORNELL UNIVERSITY PRESS
Ithaca and London

Originally published in 1966 by Rutgers University Press,
New Brunswick, New Jersey.

First Fall Creek Books edition, 2011.

PRINTED IN THE UNITED STATES OF AMERICA

To Fred W. Bogert
and
to my wife Ada
and to our children,
Elaine, Adele, Kieth

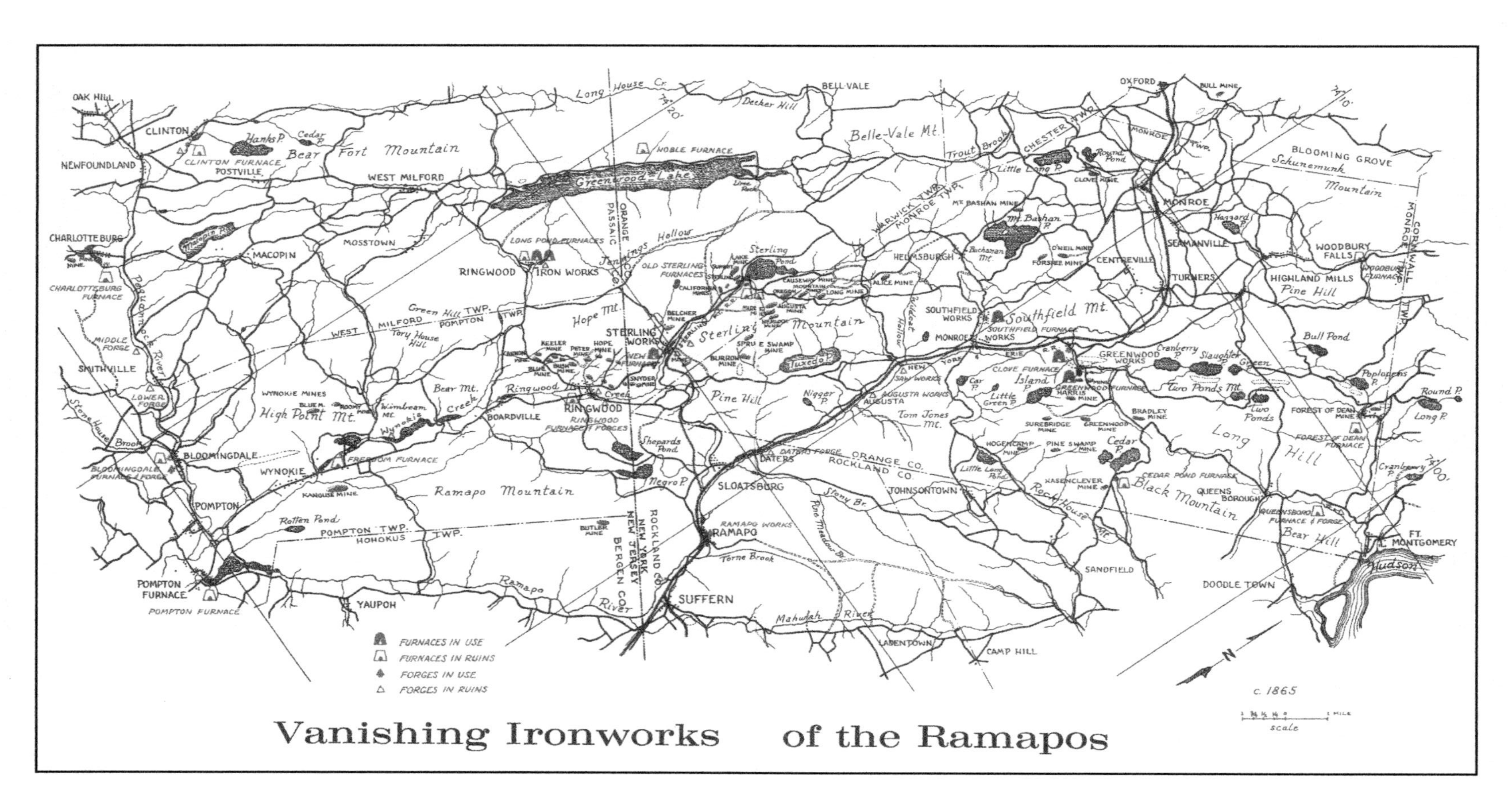
Vanishing Ironworks of the Ramapos
c. 1865
scale
FURNACES IN USE
FURNACES IN RUINS
FORGES IN USE
FORGES IN RUINS
OAK HILL
CLINTON
NEWFOUNDLAND
CLINTON FURNACE
POSTVILLE
Hanks P.
Cedar P.
Bear Fort Mountain
WEST MILFORD
Greenwood Lake
NOBLE FURNACE
Long House Cr.
Decker Hill
BELL VALE
Belle-Vale Mt.
Trout Brook
CHESTER TWP.
OXFORD
BULL MINE
MONROE TWP.
Round Pond
CLOVE MINE
Little Long P.
BLOOMING GROVE
Schunemunk Mountain
MONROE
CHARLOTTEBURG
CHARLOTTEBURG FURNACE
MACOPIN
MOSSTOWN
LONG POND FURNACES
RINGWOOD
IRON WORKS
ORANGE
PASSAIC
Jennings Hollow
OLD STERLING FURNACES
Sterling Pond
WARWICK TWP.
MONROE TWP.
MT. BASHAN MINE
Mt. Bashan P.
HELMSBURGH
Buchanan Mt.
O'NEIL MINE
FORSHEE MINE
CENTREVILLE
SEAMANVILLE
TURNERS
Haggard P.
WOODBURY FALLS
WOODBURY FURNACE
CORNWALL
HIGHLAND MILLS
Pine Hill
CALIFORNIA MINES
CAUSEWAY MINE
LONG MINE
ALICE MINE
Wildcat Hollow
Pequannock River
MIDDLE FORGE
SMITHVILLE
LOWER FORGE
Stone House Brook
WEST MILFORD
Green Hill TWP.
POMPTON TWP.
Tory House Hill
Hope Mt.
BELCHER MINE
STERLING WORKS
Sterling Mountain
SPRUCE SWAMP MINE
SOUTHFIELD WORKS
Southfield Mt.
SOUTHFIELD FURNACE
MONROE WORKS
KEELER MINE
PETER MINE
HOPE MINE
BLUE MINE
BUSH MINE
NEW FURNACE
BURRON MINE
Tuxedo P.
ERIE R.R.
NEW YORK
NEW SAW WORKS
GREENWOOD WORKS
Cranberry P.
Slaughter P.
Green
Bull Pond
Poplopens P.
Round P.
Long P.
WYNOKIE MINES
High Point Mt.
Wimbeam Mt.
Wynokie Creek
Bear Mt.
Ringwood Creek
BOARDVILLE
RINGWOOD
RINGWOOD FURNACES & FORGES
SNYDER MINE
Pine Hill
Nigger P.
AUGUSTA WORKS
AUGUSTA
Tom Jones Mt.
CLOVE FURNACE
Car P.
Island P.
Little Green P.
GREENWOOD FURNACE
HARRIS MINE
Two Ponds Mt.
Two Ponds
FOREST OF DEAN MINE
FOREST OF DEAN FURNACE
BRADLEY MINE
SUREBRIDGE MINE
GREENWOOD MINE
HOGENCAMP MINE
PINE SWAMP MINE
Cedar P.
Long Hill
BLOOMINGDALE
BLOOMINGDALE FURNACE & FORGE
WYNOKIE
FREEDOM FURNACE
Shepards Pond
DATERS FORGE
DATERS
ORANGE CO.
ROCKLAND CO.
Little Long Pond
CEDAR POND FURNACE
HASENCLEVER MINE
Black Mountain
QUEENS BOROUGH
Cranberry P.
POMPTON
KANOUSE MINE
Ramapo Mountain
Negro P.
SLOATSBURG
Stony Br.
JOHNSONTOWN
Rock House Mt.
QUEENSBORO FURNACE & FORGE
Bear Hill
FT. MONTGOMERY
Hudson
Rotten Pond
POMPTON TWP.
HOHOKUS TWP.
BUTLER MINE
ROCKLAND CO.
NEW YORK
NEW JERSEY
BERGEN CO.
RAMAPO WORKS
RAMAPO
Pine Meadow Br.
Torne Brook
SANDFIELD
DOODLE TOWN
POMPTON FURNACE
POMPTON FURNACE
YAUPOH
Ramapo River
SUFFERN
Mahwah River
LADENTOWN
CAMP HILL
N
74°20'
74°10'
74°00'

PREFACE

The story of iron-making in the Ramapos and the western part of the Hudson Highlands is a stirring chapter of early American industrial endeavor and ingenuity. It is also a tale of danger, hardship, courage, success, and failure. Above all, it depicts the struggle to convert a hard-held gift of nature into a useful resource.

For more than two hundred years, ironworks in both the Ramapo Mountains and in the neighboring Hudson Highlands made national contributions that were noteworthy. During the latter part of the Colonial era and throughout the Revolutionary and Civil Wars, as well as in the years intervening between these critical periods, essential tools for war and peacetime were turned out here. Yet the story of the ironworks—the crucial events that marked their operations and those who directed them—has only been partly told.

Though failure was frequent, in the final summary man's perseverance proved to be more durable and lasting than the solid iron ore he succeeded in wresting from the earth.

Previous books and accounts are scanty in their coverage of even the important mines, furnaces, and forges in this area. Moreover, errors of fact have been repeated without correction. Often the authors added no new material, but these early writers—Eager, Lesley, Bishop, Pearse, and Swank—made a valuable contribution. They opened up what had formerly been an unknown historical theme, revealing enough of the intensely absorbing background of the rugged ironmasters and their methods, and the rich mines and their materials, to stimulate a wish for more information about them.

During more than twenty-five years of research and on-the-spot inspection of mines, furnaces and forges, of streams, dams and mill-races, of slag piles, abandoned railroads, historic sites, and roads almost obliterated, the author has been able to uncover many new facts and to revise a number of others. He has consulted hundreds of original historical documents and journals and has interviewed men who were formerly employed at, or personally acquainted with, some of the now-vanishing ironworks. There were other valuable sources of information. In one case, an old, dirt-obscured oil painting, when cleaned, proved to be a view of the famous Southfield ironworks, enabling the author to establish the exact locations of the various buildings and to reconstruct other details.

One of the most fortunate finds of all was a group of rare photographs, taken between 1865 and 1905, of many of the old forges, furnaces, and waterwheels of the Ramapos which were still standing during those years. A discovery of this kind is difficult to match, but it *was* matched when the author came upon a number of contemporary account books which he now owns. These ledgers and journals helped fill in many gaps in information and to correct earlier writers' errors. Material—such as maps, letters, paintings, and prints—was secured through dealers, auctions, and interested friends, here and abroad, and contributed to the excitement of the research.

The author wishes to extend his sincere thanks again to Fred W. Bogert to whom this book is dedicated; to Claire K. Tholl for her fine maps and illustrations, as well as her insight and research on a number of early land transactions concerning the many ironworks properties; and to acknowledge his debt to the late Rev. Henry C. Beck who helped provide the necessary initial encouragement and confidence.

He is also most grateful to William F. Augustine, Edna R. Avera, Clarence Bartow, Gregory Bobeniether, Jane Bogert, Louise Burnett, Beatrice Cannon, Charles Capen, Jack Chard, Walter Coss, Eleanor Coxe, Frances Cronon, John T. Cunningham, Harry Dobson, Raymond F. Decker, Willard de Yoe, Howard Durie, Sanford Durland, Fred Ferber, John Fleming, Sidney Forman, Gladys Franke, Alexander Fowler, Margaret Fuller, William Gaines, Dr. Lewis Haggerty, Stanton Hammond, G. Cornelius Houn[illegible]an, Susan Roome Horton, Ernest Krauss, Otto Kuhler, Barbara Lang, Adrian C. Leiby, Edward Lenick, Richard Lenk, Rosa Livingston, Frank Malone, Pierson Mapes, Roger McDonough, Minnie May Monks, Edward Morgan, James Norman, Lewis Owen, Charles Parker, Frederick Parry, Henry Pierson, Lesley Post Richard Rhinefield,

Roland Robbins, Edna L. Royle, Salvatore Ruscelli, Edward Ryerson, Louise Ryerson, Harold Sherwood, Donald A. Sinclair, Peter Smith, Roscoe Smith, Mead Stapler, Chester Steitz, Winfred Stephens, Donald B. Stewart, Gerald Stowe, Vincent Struble, Miriam Studley, Herbert Summers, C. Edward Turse, Ethel Vreeland, Alexander Waldron, Mary and Seely Ward, Gardner Watts, Lewis West, King Weyant, Thomas Whitmore, Robert Wiggins, Nona Winters, and his wife Ada.

Thanks also go to the staffs of the American Iron & Steel Institute, City Investing Company, Cooper Union Institute, New Jersey Historical Society, New Jersey State Library, New York State Historical Association, New-York Historical Society, New York Public Library, and Rutgers University Library.

Special note is made of the thirteen remarkable photographs by Vernon Royle (1846–1934) graciously lent for publication by Mrs. Edna L. Royle and her daughter, Mrs. Edna Royle Avera.

CONTENTS

LIST OF ILLUSTRATIONS

VANISHING IRONWORKS
of the RAMAPOS

Southfield Ironworks, 1835. Detail from an oil painting by Raphael Hoyle. Owned by the author. *Photo by Frank Moratz.*

INTRODUCTION

Iron and the Ramapos

Centuries ago the Lenni Lenape Indians who lived on each side of the New Jersey-New York border gave a name to their hills which sounded like "Ramapo," but began with an "L." The Lenape language contained no "r" and the change was the white man's. Today the hills are known locally as the Ramapo "Mountains" in Orange and Rockland Counties in New York, and Passaic and Bergen Counties in New Jersey.

The Indian name was descriptive and identifying. In their language it meant "place of the slanting rock." The first white men to explore the region must have observed that the layers of rock which formed the mountains sloped, but if they did not at once observe a more important fact about them, those who came later did. To their knowledgeable eyes the gray outcroppings, with streaks of darker gray and black, indicated the presence of iron ore. It was a recognition of significant consequence.

Until the outset of the eighteenth century, most of the iron the European settlers in America were using came from across the Atlantic, largely from Britain. Its cost was high for many reasons besides the obvious one of transoceanic shipping. The mines of the Old World were old mines, and the costs of working them were high and going higher. The charcoal fuel required for smelting was becoming exorbitant.

Here in the Ramapos was part of the answer to the cost of overseas iron. This New World iron deposit consisted largely of what geologists call magnetite, a form of iron oxide with the highest known iron content; it could be smelted by familiar techniques and the necessary fuel was close at hand. The ore also contained a second sort of iron compound, hematite, more difficult to smelt and largely lost in the early refining process.

The Ramapos provided several natural advantages. The region is largely bounded by three small rivers: the Ramapo River along the southeast border of the area, the Ringwood, rising in Sterling Lake and winding southward, and the Pequannock, which rises in the eastern part of what is today Sussex County, New Jersey, and flows east-southeast into the Ramapo near Pompton Plains. Above the stream beds, the layers of ore deposit cropped out among hills thick with trees which would supply fuel for conversion into charcoal for many years to come.

The rivers, and the rains that had fed them over countless centuries, had performed another service. The rock formations which contained the ore beds had been laid down over a billion years earlier, in the epoch the geologists have named the Proterozoic. After a vast and complex series of changes, the ancient beds of ore emerged above the surface of the ocean comparatively recently, as geologists reckon time. Then the waters had gone to work, stripping away overlying deposits and carving valleys and ravines, so bringing the ancient rocks to the surface, to be seen and eventually to be mined without the necessity for difficult and costly deep working. Streams and rivers were still flowing when the white men came, as they are today, and they provided an essential for an industry of Colonial iron—adequate water power in a period which was still without the power of steam.

There was even more: the flowing waters, as they moved toward the sea, produced a final favorable element in the topography of the Ramapo region. The rivers cut into the terrain deeply, making easy gradients for the early roads. Here was a factor of special importance to an age which preceded the railroad and the canal. The ox or mule-carts of transport did not have to travel costly distances. No part of New Jersey is very far from water deep enough for shipping either light or heavy cargoes. And the water routes, in turn, led eventually to the ports and roads of all the other Colonies.

New Jersey was not the only iron producer in the New World: the same year the Pilgrims landed at Plymouth Rock a forge had been built near Jamestown; by the end of the seventeenth century Massachusetts had become the chief seat of the small iron industry

Freedom Furnace. *Photo by Vernon Royle, 1903.*

in America; and the eighteenth century saw the rapid rise of a widespread Colonial iron industry. Furnaces and forges were in operation up and down the Atlantic coast, in every Colony except Georgia, and the industry was spreading inland and westward with the frontier. Export of American iron began as early as 1718. In bars and pigs the iron was moving eastward across the ocean to Europe in ever-increasing tonnages; exports to England in 1771 were between 7,000 and 8,000 tons. By the time the Revolution broke out these Colonial furnaces were producing about 14 percent of the world's supply of iron. The Colonies were indeed producing more iron than the combined output of England and Wales.

New Jersey's major asset in the iron industry was its central location. Within the Colony itself, furthermore, there were two separate regions where furnaces could be built and not all of the iron ore lay in the northern hills. A hundred or more miles to the south, in a thinly settled area of swamps and stunted pines, there were ochre-colored deposits of another ore of iron, limonite, which the producers of bog iron had worked briefly. However, there were fewer than ten furnaces operating in South Jersey before the Revolution.

North or south, Jersey iron was easy to get at. The southern bog iron ore was to be found in surface-level deposits, and in the northern part of the state the ore-bearing strata occurred as outcroppings. In the eighteenth century, shaft mining was almost never practiced in the Colonies, including New Jersey; pickaxes and crowbars were the basic tools and manpower was therefore relatively efficient and economically practical. The eve of the Revolution found the Colonies well able to furnish their own most basic sinew of war, iron for tools, for guns, for cannon and for shot—iron enough, indeed, to forge gigantic chains to stretch across the Hudson and interdict the river to British men-of-war.

Along with the iron itself went the men who mined and smelted and wrought it. They were a special and hardy breed, and their descendants not only still bear their names but have spread into every state of the nation. What sort of men—and women—they were, where they came from, how they lived, and what became of them is a special chapter in the complex encyclopedia of American history. Their enterprises have marked the Jersey hills and changed the landscape of the state. Their story is part of the national heritage.

Furnace, Forge, and Fuel

At one time many active iron forges and furnaces were scattered throughout the Ramapo Mountains. Today a visitor to their deserted sites sees only crumbling ruins. It is difficult to reconstruct in imagination what these early ironworks looked like and how they produced the metal that Colonial farmers and Revolutionary soldiers depended upon for their very lives and livelihood. Yet, in the heyday of Ramapo iron, these almost forgotten furnaces were well-known, even famous. The men who built and owned them were often public leaders, and the miners, charcoal burners, ironsmiths, and craftsmen in metal who labored in the mines and at the hearths were independent, self-reliant citizens who took the hardships of their lives in stride.

It was an industry which demanded strenuous work, work of pioneer intensity, to mine the ore and smelt hard iron from it. Some of the story has been lost, and it is not always possible to say how accurate the existing information really is. But the documents and letters of the time do permit a reasonable reconstruction of what it was once like when furnace smokes were rising and trip-hammers banging away in the valleys of the hill country.

In those earlier days the terms forge and furnace were frequently used interchangeably; as a result, some confusion of facts as well as terms confronts anyone who investigates the story of Ramapo iron. The first smelting furnace erected in the area is believed to have been the one built by the Ogden family at Ringwood, New Jersey, in 1742. The earliest recorded forge for working iron (or "bloomery" as such forges were sometimes called) was one constructed at Pompton, New Jersey, prior to 1726. But the exact year it was built and the date when its actual operation began are unknown.

Furnaces can be described simply as truncated pyramids of stone and brick, and they were usually located at the foot of a hillside. From a point part way up the hillside, a bridge or loading platform was built across to the top of the furnace stack. Progressive layers of fuel, ore broken up into fist-sized chunks, and limestone for flux, were taken across the bridge by baskets or crude pushcarts and emptied down the open stack of the furnace. This charging was the first step in preparing a furnace for firing. By modern standards, these early iron-working furnaces, or forges, were unimpressive in size. The earlier furnaces, using charcoal as a fuel, were seldom more than 25 feet square at their bases and ranged from 22 to about 30 feet in height. The later furnaces, built during the 1840–60 period, used Pennsylvania anthracite and rose as high as 55 feet. They averaged some 35 feet across the square base.

The egg-shaped interiors of the earlier furnaces were lined with sandstone or slate against the intense heat generated when the

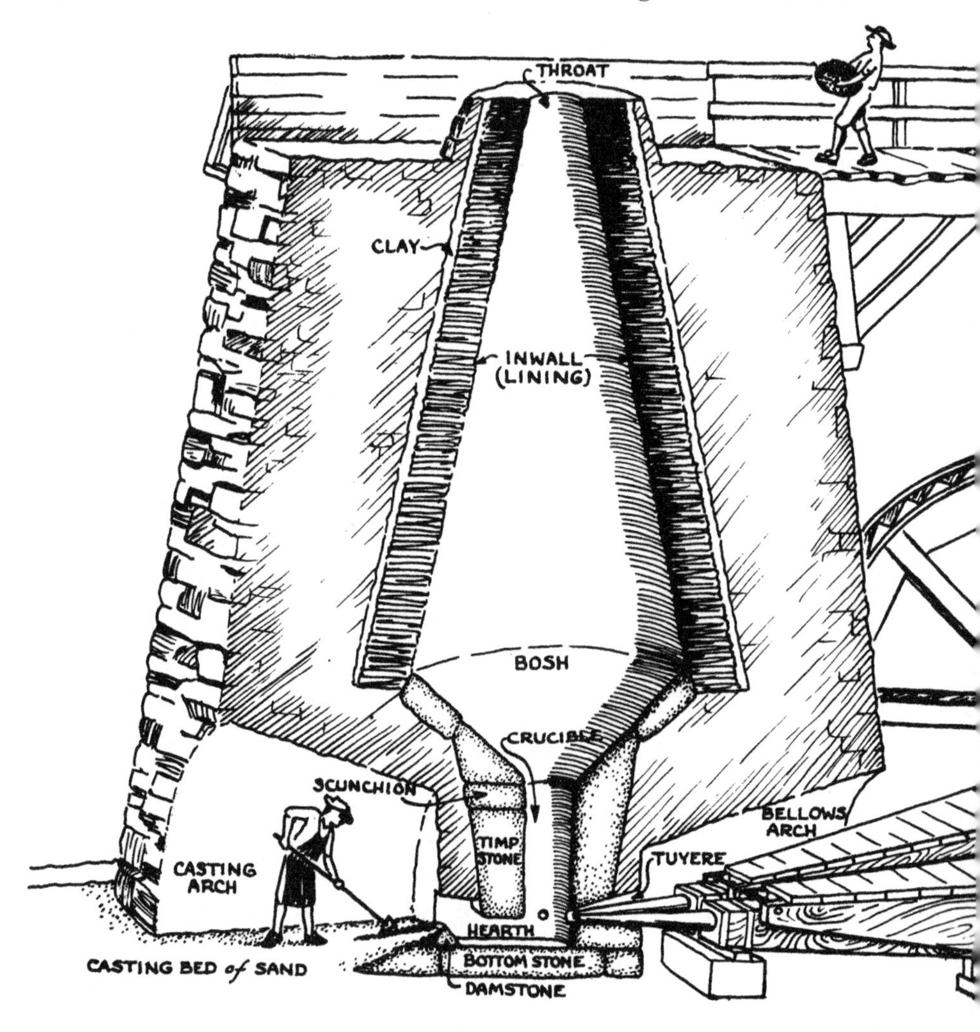

furnace was operating. The later ones employed firebrick. The blast, or forced stream of air which intensified combustion, was supplied by two or more large bellows, operated alternately by cams on revolving shafts, which in turn were powered by waterwheels. Many of these wheels were 15 feet or more in diameter, and an ample supply of water was needed to turn them. For this reason, furnaces were invariably located near good-sized streams or even rivers. Some of the early ironmakers constructed reservoirs upstream from their furnaces and forges to forestall lack of water during a pro-

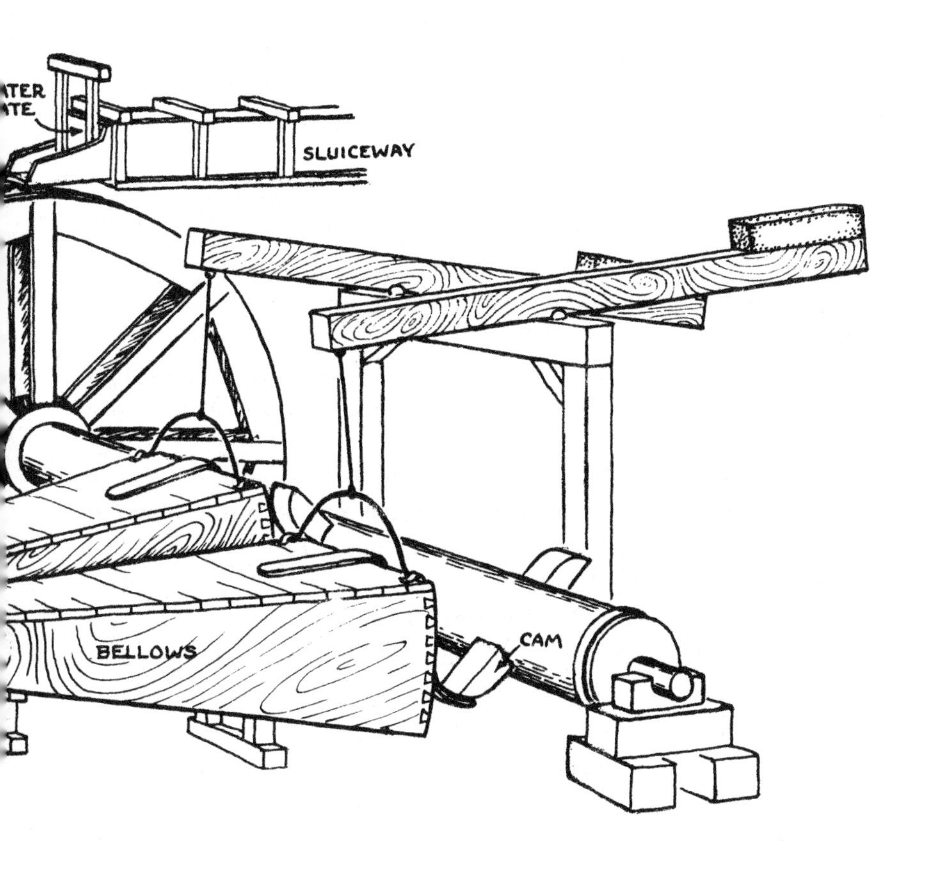

Molten iron flowing from the furnace into the pig beds.

tracted dry season, and built wooden roofs to shelter the water-wheels from winter weather.

Once filled to its proper level, the furnace was ignited and the process of smelting the iron ore began. Intensified by the air blast, the heat of the layers of burning fuel gradually melted the ore. As the process progressed, the iron, denser and heavier than the ore, trickled to the bottom, or hearth, of the furnace. The residue, or slag, floated on top of the molten iron and was drawn off through an opening—the cinder notch—just above the damstone at the base of the furnace.

When the tapping hole in the casting arch was unstopped, the molten iron flowed forth. It ran into a trough and on into a channel (the sow) dug in the sand of the casting floor. From the sides of the sow a series of smaller troughs extended. The liquid iron flowed into these laterals as well. The pattern formed by the casting bed resembled the outline of a mother pig feeding her sucklings, and suggested the term commonly used by the ironmakers and the industry—"pigs"—to describe iron cast in this fashion. Sometimes additional channels ran from the sow into molds for firebacks or other cast iron products, but pigs were the principal output.

In addition to the casting arch, the early furnaces had another arch, on a second of its sides, to provide entry for the bellows blast which passed through tubes called tuyère (pronounced twee-air) pipes. Later furnaces had a third arch, also constructed to admit tuyère pipes. Still later, a fourth arch was added to provide blast from three sides. The arches generally approximated 12 feet in width and rose as much as 15 feet from the ground to their apexes.

After the iron pigs had cooled in the casting bed, they were taken to the forge. Here they were reheated to softness and shaped into lumpy masses which were then pounded by a massive water-driven trip-hammer to reduce the impurities in the iron. Next, the iron was reheated in the forge, brought back again for hammering, and finally shaped into bars or forms, called "anconies," resembling dumbbells.

An important part of early iron manufacture was the production of charcoal for fuel. At least until the beginning of the nineteenth century, and even for a short time thereafter, ironmasters relied on charcoal as the principal means of fueling their furnaces. Producing charcoal, like the rest of the process of iron-making, was a laborious task even though in the hilly country of the Ramapos wooded terrain was always relatively close at hand. Gangs of wood-

Shaping an ancony at the forge. Denis Diderot, *Encyclopédie*, 1765.

choppers first felled suitable trees and after the trees had been cut into proper lengths, trimmed the logs and carted them to a charcoal-burning area. Here they were stacked upright in a cone-shaped pile on dry, level, and sheltered ground—then covered by earth and damp leaves.

The blanketed cone of wood was set on fire by igniting chips and leaves and dropping them down the small flue or chimney left in the process of stacking the logs. Firings were usually carried on from early May until late in October, thus avoiding the problems of strong winter and spring winds, and other unfavorable weather conditions. Weather, indeed, dictated the entire cycle of charcoal burning. The trees themselves were cut down in the coldest months, after their sap had retreated to their roots. Wood cut in the winter months is harder and dries faster than at any other season of the year, and when burned, it produces stronger and more solid charcoal as the sap of deciduous trees, going down into the roots for winter, leaves behind in the fibers of trunk and branches a large proportion of the soluble chemicals.

The firings called for skill and experience. Small openings had to be provided in the sides of the cone of wood to supplement the central flue in providing a proper draft, and had to be adjusted to changing winds. Since from three to ten days were required to char a large cone properly, depending partly on the weather and partly on the kind of wood being charred, the whole process called for constant and skilful surveillance, and burners often camped out in rough shelters near the smoldering cone to maintain a round-the-clock watch on the operation. The charcoal end-product of all their work and care was nearly pure carbon which, after cooling, was heaped in small piles and taken to the furnaces as it was needed.

Much more than just iron ore—furnaces, forges, charcoal, sandstone, limestone, and heavy labor—was necessary for the successful operation of an early ironworks. A whole catalog of supplementary goods and services was needed to provide and maintain the wide variety of tools, mauls, sledges, utensils, and other implements necessary to the process of iron-making itself, and to the human needs of the workers who manned the works and their families. Inevitably, communities grew up in the vicinities of the furnaces and forges. There were stores, offices, gristmills, blacksmith shops, carpenter shops, homes, schools, churches, and eventually even railroads to provide transportation to and from the sites of the early ironworks. From the beginning, such enterprises demanded expert management and coordination, but occasionally administrators were not equal to the problems. At least once, an ironworks well-known in the Ramapos failed because of too ambitious an outlay of resources to meet the demands, social and technical, of the enterprise.

A charcoal burner's hut. Westbrook Valley, West Milford, New Jersey. *Photo by Vernon Royle, 1895.*

I

FORGES AND FURNACES IN NORTHERN NEW JERSEY

Peter Hasenclever
1716–1793

PETER HASENCLEVER AND THE AMERICAN COMPANY

Among the colorful and enterprising figures connected with the Colonial iron industry in the Ramapo Mountains, Peter Hasenclever is the most remarkable one of all. He was the first to recognize the possibilities in volume production offered by the mines in this area and to attempt to take advantage of it. At a time when others were thinking in terms of a single forge and furnace, Hasenclever indulged in large-scale planning and enthusiastically set out to establish an iron empire in America.

He was born in Remscheid, Germany, which was then in the Duchy of Westphalia, on November 24, 1716. At fourteen, he worked long hours in a steel mill, an experience that was to prove valuable when he operated Ringwood and built the Long Pond and Charlotteburg ironworks. Later, he sold cloth and needles, traveling all over the European continent as the representative of his cousin, a manufacturer of sewing materials. In addition to acquiring a useful knowledge of the French language, he developed the business sense that later helped him to convince his backers and partners that the success of iron manufacturing in America was inevitable.

Arriving in London during the summer months of 1763, he established residence for himself and his family and became a British citizen. He formed new business connections and became a partner in the commercial firm of Hasenclever, Seton and Crofts, whose capital funds—20,000 pounds sterling—were earmarked for investments. The opportunities for investing money in England or on the European continent were plentiful and Hasenclever's experience in these areas would give any venture a more than even chance of success. But he turned away from these more familiar areas and looked instead to the west. "Having examined, when on travels," he explained:

> the most considerable manufactures in England, Germany, France, Sweden and Spain; being most theoretically acquainted with the nature of the steel, iron and potash manufactures, and the culture of hemp, flax and madder and being informed that there were in North America rich steel and iron mines, with inexhaustible woods, full of timber for coal—an article exceeding scarce and expensive in Europe. . . . [I was] animated to quit England for a year, or 18 months at farthest, in order to go to America to make such establishments as might enable me to enjoy these advantages. . . .[1]

Although he had never been to America and did not know whether he would be able to find an ironworks for sale, or property with all of the ore, forests, and waterways necessary for the manufacture of iron, Hasenclever sent to Germany for skilled laborers. Here, in the heart of the best iron and steel-making region in Europe, he hired forgemen, furnacemen, charcoal burners, miners, masons, carpenters, and other experienced help. With characteristic attention to detail, he even arranged transportation for these workmen and their families to the American Colonies.

His self-assurance and faith in the success of this undertaking must have aroused feelings of both confidence and bewilderment in the minds of his newly acquired partners. Hasenclever undoubtedly transmitted his enthusiasm for the venture to some of his friends as well, because early in 1764 they agreed to spend from 10,000 to 40,000 pounds, as circumstances might require, to complete the enterprise.

Having arranged with his partners, Seton and Crofts, for the operation of the London firm during his absence, and with this promise of support from his friends, Hasenclever sailed from England April 20, 1764, arriving six weeks later in New York. From an advertisement in a New York newspaper, published prior to his departure from London, he learned that the Ringwood Ironworks Estate, which he described as "a decayed Ironworks," was for sale because of the lack of "able workmen." Eager to inspect it, he traveled through fertile fields with well-kept Dutch farms to the heavily wooded hills of the Ramapo Mountains where the Ringwood works were located. In spite of the "decayed" condition of the equipment and property, Hasenclever lost little time in making up his mind to buy it. On July 5, 1764, he purchased the Ringwood Ironworks Estate for 5,000 pounds sterling from David Ogden, Sr., David Ogden, Jr., John Ogden, Uzal Ogden, Samuel Gouvernour and Nicholas Gouvernour.[2]

Hasenclever repaired the rundown works and "made Iron" almost immediately. He now became convinced that he needed to add other properties to expand his dominion beyond the Ringwood ironworks. Confident of his ability for making a success of any mining venture, Hasenclever made plans to build a total of 5 furnaces and 7 forges with 12 hammers and 23 fires—a tremendous undertaking for those times. He estimated that the ore in the Ramapo Mountains was so plentiful that each furnace should be able to make 700 tons of pigs a year, about 3,500 tons. Computing the possible yield from future as well as present forge fires, he was sure he could turn out at least 1,250 tons of bar iron a year. He estimated the profits from this large annual volume at over 13,000 pounds sterling, a huge sum indeed in the eighteenth century.

Hasenclever, now with many bigger things in mind, "acquired by degrees [by patent], for account of the American Company, upwards of 50,000 acres of land, for the use of the Iron works, and for the planting of Hemp, Flax and Madder." Upon some of these lands, he built the Charlotteburg and Long Pond works in New Jersey and the Cortlandt and Cedar Ponds works in New York. Because the quality of the ore found at Cedar Ponds did not come up to expectations, the furnace planned at this site was never completed. Later he was forced to halt operations at Cortlandt furnace for the same reason.

By the autumn of 1764, many of the workmen hired in Germany by Hasenclever began to arrive with their families. Within a year after the first group had landed in America, more than 500 of them had been transported to the various properties, involving considerable expense for the firm of Hasenclever, Seton and Crofts. The ambitious German, at the same time, was buying "122 horses, 214 draft oxen and 51 cows, and a vast number of implements for the works . . ." More funds were laid out for building roads, dams, reservoirs, and various other structures. Within a two-year period, Hasenclever had spent money so freely he was dangerously close to the limits of the capital pledged by his firm—and had shown little return on the investment. Hasenclever claimed that "the working of so many mines was not only a very expensive, but also a laborious and vexatious work."

While he was busily engaged in creating this enterprise in the Ramapos, word reached Hasenclever that one of his partners, Andrew Seton, was indulging in a spending spree of his own and he decided to investigate this report in person. Before sailing to England, Hasenclever also received the surprising news that he had acquired some new partners, but at this time he did not know the extent of their financial interests. This new group, as he later discovered, was calling itself the "American Company" and occasionally the "London Company." But just how the tie-in came about and how the new company was related to the original firm of Hasenclever, Seton and Crofts (apparently never legally dissolved), is not clear.

Hasenclever left New York in November of 1766, and upon arriving in London lost little time in finding out about his new partners. They included several prominent figures in the military, political, and mercantile circles of the time. Among them were Major General David Greeme, a Deputy Secretary of the Admiralty named George Jackson, John Elves, and Richard Willis. At a meeting of the partners on May 11, 1767, Hasenclever gave them a detailed report of the American properties. His plans for producing great quantities of iron at substantial profits must have convinced them, at least for the time, of the promising opportunities in the Ramapo Mountain area. To show its confidence in him the American Company engaged Hasenclever to return to the Ramapos and carry on the operations as its manager. A new deed of partnership was drawn up and signed by the new partners and Hasenclever.

Prior to the meeting, Hasenclever had his suspicions about Andrew Seton confirmed by the third partner, Charles Crofts. According to Crofts, their former colleague was now completely bankrupt. Hasenclever had narrowly missed having his dreams shattered by the rash acts of the spendthrift Seton. But with the newly-signed contract safe in his pocket, he could be assured of solid backing for continuing his task of building an iron empire.

Hasenclever sailed from London on June 1, 1767, disembarking two months later at New York after a slow passage. He eagerly set out for his ironworks to learn how things had progressed during his absence. To his great distress, he found that while he had been away work had ceased entirely. Instead of nearing completion, the furnaces, forges, and other key buildings had actually deteriorated. Hoping to recover lost time, Hasenclever made plans to redouble the efforts of his workmen. But a greater blow was soon to follow.

Just forty-six days after his return from the American Company meeting in London, Jeston Humfray arrived unexpectedly at Hasenclever's impressive estate at Ringwood. He presented orders from the American Company that he was to take over the management of all the company properties in the New York-New Jersey area as of October 1, 1767.[3] For reasons still unknown, the new partners had changed their minds soon after Hasenclever's departure.

This blow must have stunned Hasenclever. However, not long afterward, he had recovered sufficiently to criticize severely the abilities of the man the new partners had chosen to succeed him. Hasenclever is quoted as saying that "If they had enquired all over Europe, they never could have found a more ignorant, obstinate Manager." He also attacked their selection of Humfray on the grounds that the American Company's new candidate, never before having managed an ironworks, was utterly unqualified to be placed in charge of an enterprise of this magnitude. Furthermore, Hasenclever backed up this statement with pointed examples of mismanagement with which he charged Humfray:

> He discharged the most expert workmen, and engaged ignorant people (who burnt both Coal and Iron) in their place; he burst a new furnace by his ignorance; he dug up pipes, which I had laid underground to secure them from the frost in Winter time, by which the water was conducted on the furnace-wheels; these pipes he laid on a scaffold; the first cold weather froze them, and the furnace was extinguished for want of water to turn the wheels; he pulled down a new overshot forge, and built it undershot. Two of the best new forges and several houses were burnt by neglect; he kept about an hundred superfluous labourers; . . . I now plainly saw, that instead of having a dividend of 7 per cent. profit, amounting to 420 [pounds] on my six shares, I should certainly be obliged to pay such a sum annually for losses. . . .[4]

Afraid of Hasenclever's reaction to his removal, the American Company petitioned Lord Hillsborough, Secretary of State for the American Colonies, to order the governors of New York and New Jersey to protect the company properties from any attempts he might make to destroy or sabotage the works. Hasenclever, interpreting this move as one intended to injure his credit and slander his character, counterpetitioned William Franklin, the Royal Governor of New Jersey, within whose province most of the ironworks lay, to examine and report on the condition of the properties. Hasenclever's plea was made not only to the chief executive of the royal province in which he lived, but also to a personal friend. Franklin had been associated with Hasenclever in several public-spirited ventures including the establishment in 1766 of Queens College, now Rutgers, the State University of New Jersey. Hasenclever's circle of acquaintances reached into almost every part of the two provinces. He negotiated land transactions in the Mohawk Valley area of New York with the noted Sir William Johnson, His Majesty's Superintendent of Indian Affairs in North America, a baronet of great distinction throughout the Colonies. He was on writing terms with General Thomas Gage, Commander in Chief of His Majesty's Forces in North America from 1763 to 1775. Still another friend was the well-known Hudson River landowner and figure of political prominence, Robert Livingston.

On June 27, 1768, Governor Franklin named a committee of four leading citizens to investigate and report on the condition of the American Company's various holdings. They were William Alexander, the self-styled Earl of Stirling and part owner of the prosperous Hibernia ironworks in neighboring Morris County, New Jersey; Captain James Gray, another fellow ironmaster from Little Falls, not far from the Ringwood mines; Colonel John Schuyler, proprietor of the famous Schuyler copper mine near Second River (now Belleville), New Jersey; and Major Theunis Dey of "Lower Preakness." [5]

Only five days after their appointment to the Franklin Committee, these men arrived at the Charlotteburg works on the west branch of the Pequannock River, just west of the Ramapos. After looking over the state of the works there, they proceeded "about thirteen miles to Ringwood, situate on a more northerly branch of the Pequanock River which is called the Ringwood River and is in Bergen County." The third and last of the Hasenclever-constructed works to be visited by this inspection team was "three miles south westwards . . . the Long Pond Works, which are situated on a stream which issues out of the Long Pond [present Greenwood Lake] which is astride the New York-New Jersey boundary in the northwestern portion of New Jersey and falls into the Ringwood River, about four miles below the furnace." By July 8, 1768, a mere eleven days after starting the investigation, the committee finished its report and sent it on to Governor Franklin.

Despite the short time it took, the report presents an amazingly complete picture and appraisal of the three chief ironworks Hasenclever had erected, providing ample proof of the accomplishments brought about within a relatively brief span of time. Not only are the smallest details regarding the sites of dams, forges, and furnaces spelled out, but also the yearly production potentials each works could be reasonably expected to yield. Careful enumeration of the numbers and types of buildings and their particular purposes are also listed.

It is obvious that the examining committee not only approved, but even applauded the ingenuity and resourcefulness of the German ironmaster. The report says:

> He is also the first we know of, who has rendered the old cinder-beds of the furnaces useful and profitable; for, at Ringwood, he has erected a stamping-mill to separate the waste iron from the cinders, by which means some hundred tons of small iron have and may be obtained, which is as good as the best pig iron. He has also made a great improvement in the construction of the furnaces, by building the in-walls of slate; which, by the experience he has already had of it, will, in all probability, last many years; whereas the stones commonly made use of for that purpose, seldom stood longer than a year or two, and would often fail in the middle of a blast.

After listing some of the innovations introduced by Hasenclever and "worth attention," the committee summed up:

> On the whole, it is a matter of surprise to us, to see such a number of great works of various kinds at different places, executed in so compleat and masterly a manner, under the direction of one person,

> in a new uninhabited country, within the short space of time that has elapsed since Mr. Hasenclever first began them; and we must here observe, that the buildings of all kinds seem to us to be commodiously contrived, all of them useful and none of them unnecessary.

Apparently the American Company decided to honor the bills of credit and debts incurred by Hasenclever in its behalf, for by the first of the following year the company paid Hasenclever 13,177 pounds, but still owed him more than 14,000.

The first of a confusing series of suits and countersuits in England followed, not to be settled until after Hasenclever's death twenty-four years later. In 1769, Hasenclever left for London, never to return, but legal disputes still went on and on. In 1770, several of the trustees of the American Company succeeded in preventing Hasenclever from obtaining a certificate to "remain free and engage in trade." The ironmaster's dream of a great iron empire was nearing its conclusion.

It is easy to see why Hasenclever was a controversial figure among his contemporaries. Although credit is universally given to his capabilities, charges were leveled at his poor judgment in his overfree use of capital funds. He was inclined, his critics said, to allow his ambitions to outdistance his partners' money. His tremendous land holdings—properties in Orange County, New York, and in New Jersey's Bergen and Morris Counties; 18,000 acres in Herkimer County, New York; a pearly ash plant in the German Flats region of the Mohawk Valley; other lands around Lake Champlain and in Nova Scotia, Canada—were cited as evidence of one person's acquiring more property than he could profitably handle. To some of the people who wrote about him, he was a visionary whose schemes were impractical; to others he was a sound, experienced businessman whose knowledge of iron manufacture and eighteenth-century mercantilism could have created a strong and profitable industry if only the American Company had kept faith with him.

Much of the derogation of Peter Hasenclever's reputation may be traced to letters written by Robert Erskine, which appeared in the *New-York Gazette and the Weekly Mercury* in 1773. These letters were an attempt to discredit a pamphlet written and circulated by Hasenclever earlier that year in London by which he hoped to clear his good name from what he considered slurs cast on his character and judgment. Hasenclever's pamphlet said in part:

> There is, perhaps, no branch of business in the world so subject to disappointment as mines; sometimes signs of the greatest riches appear as infallible, and, in a short time all disappears again. The accidents I encountered, and the anxiety and vexation I suffered, on account of the mines, are beyond description; these disappointments and the damages occasioned by the floods, were losses, which, in their consequences, doubled the expenses.[6]

Erskine had arrived in America in 1771, four years after Hasenclever's dismissal, and had become manager of the American Company properties. He evidently felt a compulsion to show loyalty to his new employers by writing against his predecessor: "The confidence reposed in me, by the proprietors of Hasenclever's iron works, I have no doubt, must sufficiently apologize for my endeavors to do them justice by publishing my observations and sentiments."[7]

Although he admitted that he "never saw that gentleman," Erskine attacked Hasenclever with a severity that seemed like personal animosity. Erskine was largely responsible for blackening Hasenclever's reputation, particularly in regard to his ability. "To Mr. Hasenclever himself then we are indebted for the history of the rise, the progress and the fate of an undertaking projected, commenced and carried on with all the rapidity, the imprudence and the profusion that the most sanguine schemer could suggest. . . ."

Despite Hasenclever's craving for the luxuries of life: large handsome estates with many liveried servants; well-turned out horses and carriages complete with footmen; brass bands at dinner every evening; and even the self-bestowed title of "Baron," he had no intention of deliberate fraud. James Rivington, the New York printer and bookseller, wrote Sir William Johnson on September 16, 1769, that Hasenclever's "fate was regretted for he was honest and well loved."[8]

Perhaps Peter Hasenclever was, as some have called him, a man "ahead of his time," and he sensed this in himself. On January 6, 1768, he wrote Sir William Johnson:

This country is not yet ripe for manufactures. Labor is too high—too much land to be settled. To erect fabrics [factories] is to ruin the landed interest. The country people must resort to towns and the land will lie waste and incult. Fabrics should not be established, then in countries where are [not?] more people than what can be employed in agriculture, and therefore I think the present zeal to establish manufactures is premature.[9]

Hasenclever coat of arms. Courtesy of R. Curt Hasenclever.

The frustrating law suits dragged on and Hasenclever left London for Germany in 1773, settling in the community of Landeshut, Silesia. Frederick the Great entrusted him with the task of improving the linen business in the German states. Hasenclever seems to have been most successful in this endeavor, and he remained there in the linen trade until his death on June 13, 1793. Ironically, less than seven months after his death, his family received a letter informing them that the long-drawn-out court proceedings in England had finally come to an end with a decision in favor of Hasenclever, Seton and Crofts for approximately 72,000 pounds sterling plus interest at the rate of 5 percent for twenty-four years. No one was ever to collect: The American Company was in bankruptcy. Vindication had come too late.

Predominant in Peter Hasenclever's character was tremendous courage, indomitable spirit, and what is defined today as "know how." These qualities, despite numerous obstacles, enabled Hasenclever to establish the first large-scale iron operation in the Colonial wilderness—a mighty forerunner of today's vital steel industry.

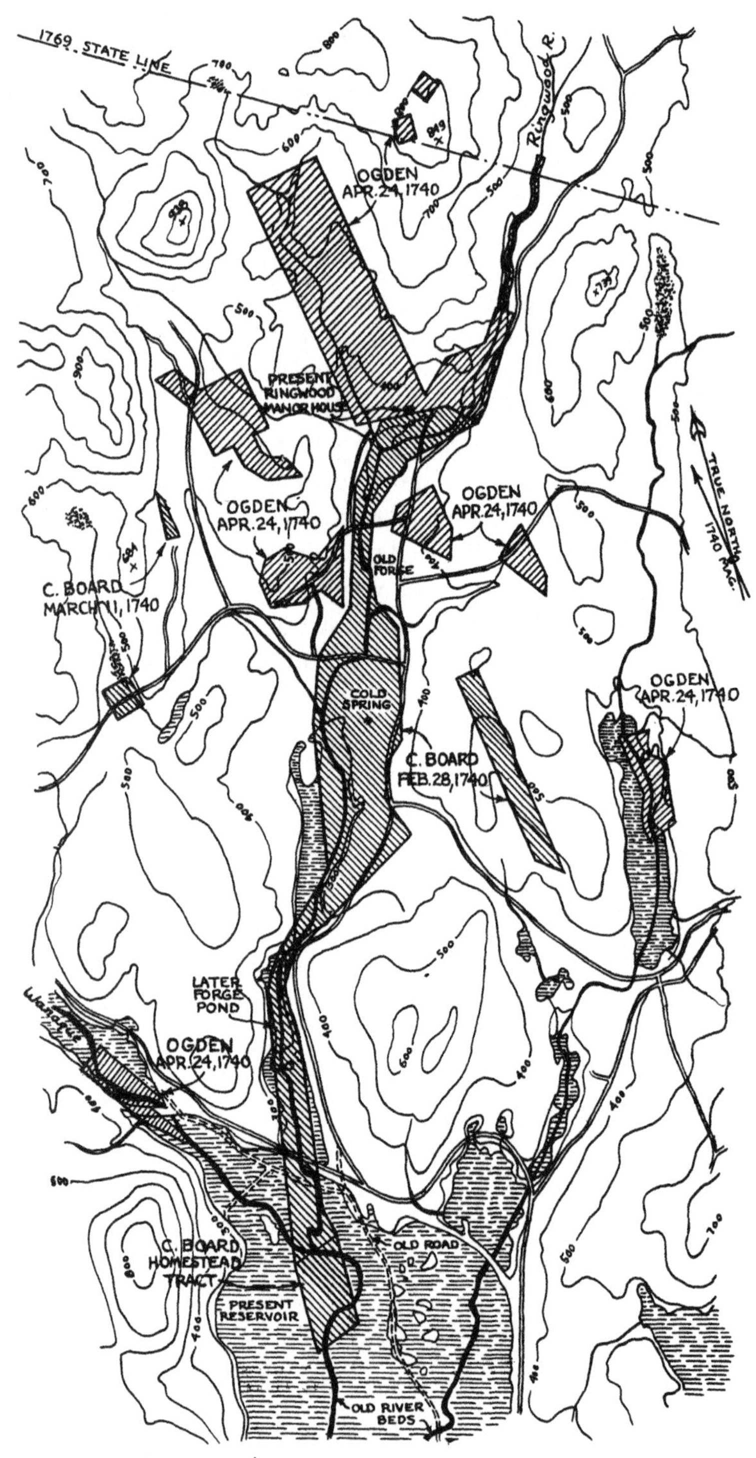

C. BOARD & OGDENS' 1740 LAND ACQUISITIONS AT RINGWOOD
overlaid on current map

RINGWOOD IRONWORKS

In 1736, more than twenty-five years before Peter Hasenclever bought the Ringwood ironworks, Cornelius Board, who had been in America only six years, explored the Ringwood River in the uninhabited hills of northern New Jersey.[1] Board, an Englishman of Welsh descent, was looking for precious metals—gold and silver, as well as copper—which he believed might be found in the land stretching over the boundary from New Jersey into New York. Although he failed to find these metals, Board did notice frequent outcroppings of iron ore.

That fall, Board and a partner, Timothy Ward, purchased two tracts of land in this area which included a sparkling blue lake known today as Sterling. Shortly afterward, they built a bloomery at the foot of the lake, probably the first ironworks in the state of New York, and started manufacturing iron.[2] This land acquired by Board and Ward was the first of a series of purchases that extended, for the most part, from the Eagle Valley area on the south to the head of Sterling Lake on the north.

Board sold out his interest in the Sterling ironworks in 1740 to Ward and moved to the Ringwood area where, on February 28, he had purchased three tracts of land from the Proprietors of East Jersey. The largest of the three pieces of land contained 246.76 acres and reached all the way from the junction of the Ringwood and Long Pond rivers to the vicinity of the millpond and manor house at Ringwood. There he promptly erected a forge about a half mile south of the manor house, on what is now Cooper Union property.

He built his home on the second tract, 28.13 acres on the southern boundary of the first tract. The third parcel, slightly more than 30 acres, lay on a hill east of the first and largest tract.

In less than two weeks' time after acquiring the three tracts,

Board was busy securing even more property. On March 11, 1740, he purchased two pieces, totaling ten acres, lying to the west of his largest tract. Although small in size, they were rich in iron deposits. By now Board had acquired extensive holdings at Ringwood, which early deeds often referred to as "Board's Plantation."

Curiously enough, on April 15, 1740, he sold sixteen acres of his recently acquired property to the Ogden family of Newark, New Jersey, for 63 pounds sterling, though he had owned the land for less than two months. This particular tract was located just to the north of the junction of the Long Pond and Ringwood rivers, and probably the ironworks which became known as the "lower forge" was built soon after. The Ogdens also bought land from the Proprietors of East Jersey that month.

There were nine tracts involved in this transaction, the largest being 165.78 acres lying north of the manor house. The greater part of this was on the west side of Long Hill, but a portion of it extended to the Ringwood River, with a very narrow strip following north along the river beyond the present state line. The small red house, which still stands, and the millpond are both there. Five of the other tracts were located along the Long Pond River near Board's home. Another was situated to the northeast of the forge site on the Cooper Union property. This was a 40 acre tract upon which the Ogdens were soon to erect a furnace. The other two tracts, where some mining was done, were small, less than three acres each, and north of the manor house on Long Hill.

With the acquisition of these properties, the Ogdens formed the famous Ringwood Company and in 1742 built the first blast furnace to be erected there, thus becoming the first volume producers of iron. Very little is known about the operations at the Ringwood Company's ironworks during the twenty year period through 1760, except that the Ogdens did acquire a new partner, Nicholas Gouvernour, who advertised in the *Pennsylvania Journal*, May 24, 1759, the sale of refined and stamped bar iron, and some "Share" molds made at Ringwood. Gouvernour claimed in his advertisement that the quality of the iron was so excellent that it was most suitable for shipbuilding and "country work." This iron was judged the equal of the best Swedish iron which had a worldwide reputation for the finest quality.

In 1762, the Ogdens built a new furnace at Ringwood. Two stones from this early Colonial stack still survive and have the year "1762" marked on them. One can be seen on the ground at the southside of the manor house wing at Ringwood. The other was discovered in the cellar of the house, coincidentally, just two hundred years after the building of the furnace.[3] This first mentioned stone has the initials "D. B." and the latter bears the initials "H. M." It is possible that "D. B." stands for David Board, a son of Cornelius. Although it is more than likely that the stones forming the cascade by the driveway are also from the same stack, these two stones are the only definite remaining traces of this old Ogden furnace that helped make the name of the Ringwood Company so well known in Colonial America.

According to an account book of the Ringwood Company kept by the Ogdens from November 14, 1760 through April 11, 1764, David Ogden, Sr., David Ogden, Jr., and Nicholas Gouvernour "settled up and closed" the Ringwood works on December 10, 1763. However, both Ogdens and Gouvernour must have continued to purchase land, either for their own use or for company purposes, as they are listed as buying six-and-a-half acres for 6 pounds, 10 shillings on February 1, 1764, from Joseph Board, son of the pioneer in this area.[4] The same day they also acquired from the same owner "a tract of land scituate [sic] lying and being at Ringwood, near the Old Forge and dwelling house of Walter Erwin." [5]

Apparently the settling up and closing described in the Ogdens' account book was a preliminary to offering the Ringwood Company works for sale on the open market. At least an advertisement in the *New-York Mercury,* March 5, 1764, seems to point to this. The properties and access to shipping facilities are clearly described:

To Be Sold

> A new well built furnace; good iron mines near the same; two forges—one with three, the other with two fires—a saw mill, several dwelling houses and coal houses; and several tracts of land adjoining; and forges are situated on a good stream 28 miles from Acquakanung landing and 36 from Newark. Whoever inclines to purchase the same may apply to Nicholas Gouvernour in New York; or to David Ogden, Sr., Samuel Gouvernour and David Ogden, Jr., at Newark who will agree for the same.

No doubt it was this advertisement that attracted the attention of Peter Hasenclever when he was eagerly seeking iron manufacturing facilities. The deed which Hasenclever received when he purchased the Ringwood Ironworks Estate for 5,000 pounds sterling on July 5, 1764, listed not only the "Furnace, two forges and several dwell-

Drawing of a furnace from Ringwood's cash book, owned by The New Jersey Historical Society. This was found between entries dated Dec.

ing houses" cited in the advertisement, but also refers to Ward's earlier forge and the "old Forge." The deed for the sale was signed by the Ogdens, father and son, and the Gouvernours. John and Uzal Ogden deeded their share in the properties to Hasenclever on the same day, but in a separate conveyance.

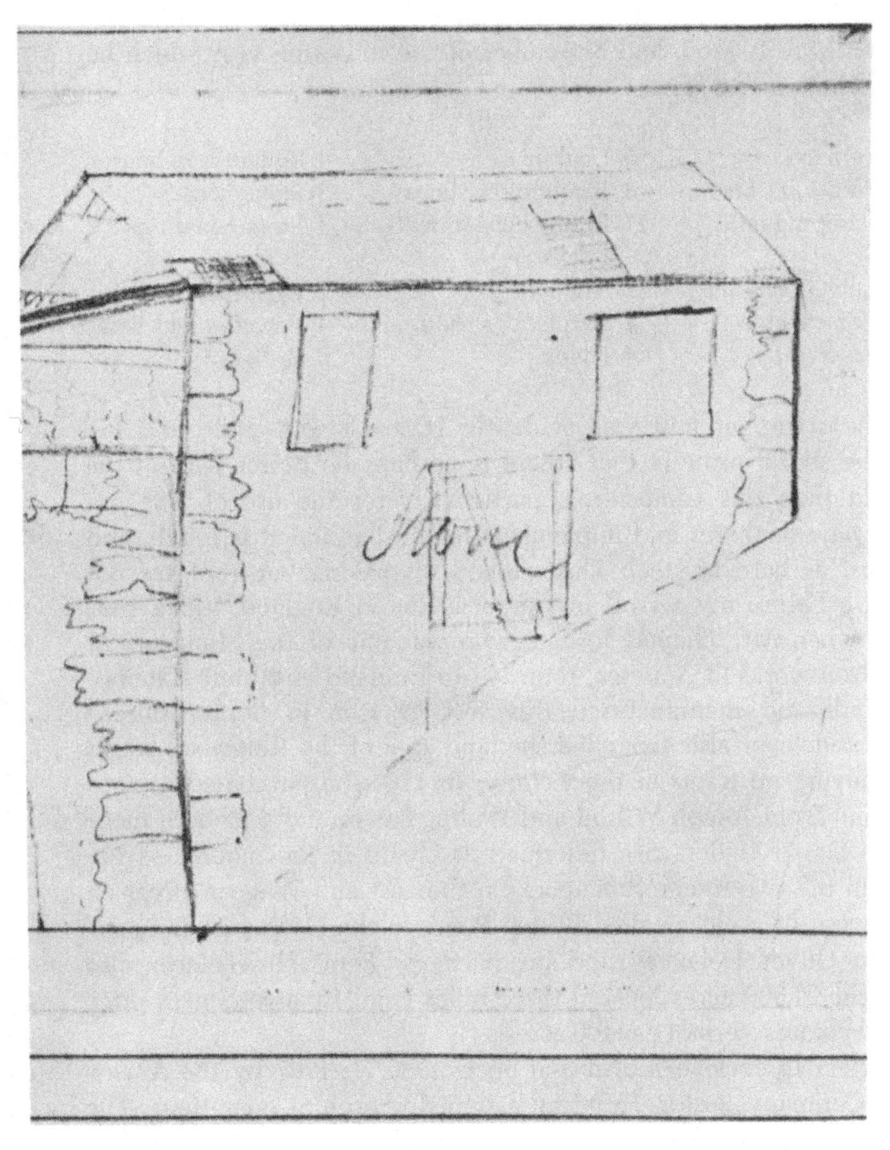

775 and Jan. 1776. The stones bearing the dates and initials are all that emain of the furnace.

Hasenclever referred to his newly-acquired works as being "decayed," yet four months after he bought them, the Ringwood Company works were turning out iron. Despite the fact that the Ogdens had built a new furnace just two years before, Hasenclever was obviously displeased with conditions as he found them. He claimed to have built or rebuilt a large number of necessary facilities between May 1, 1765, and November of the following year which he listed in itemized form:

1 Furnace	5 Coal-houses	4 Square Log-houses
4 Forges, 11 fires	3 Blacksmiths shops	3 Stone-houses
1 Stamping-mill	17 Frame-houses, with bricks	1 Store-house
25 Colliers Houses	1 Saw-mill	1 Grist-mill
1 Horse-stable	1 Carpenter's shop	4 Barracks and barns
1 Reservoir	4 Ponds	2 Bridges [6]

The stamping mill was evidently Hasenclever's pride and joy for he stated proudly that "there is perhaps no better Iron in the world than this Cinder-iron, particularly for the use of waggon-tire, plough shares and implements for husbandry; it is tough, and almost as hard as steel. This method of making bar-iron was not known before my arrival in America, nor in England, 'till a year ago, when Mr. Thomas Jordan, who was one of the Managers of my Iron-works in America, returned to England and built a stamping-mill and manufactured this sort of Iron in Staffordshire."[7]

Hasenclever also expanded the land area of the Ringwood works by buying up tracts in the vicinity. In 1764, he purchased a piece of land from Joseph Wilson and Walter Erwin and a 68-acre piece from David Ogden, Sr., described as "lying in the mountains between the two rivers, Romapock on the east and Wancue River on the west, at a place called Rotten Pond, in the County of Bergen." From Oliver Delancey and his partners, Peter Hasenclever also bought 10,000 acres located three miles from Ringwood at a price of 30 pounds sterling per 100 acres.

After Hasenclever's dismissal on October 1, 1767 by the American Company, Jeston Humfray assumed charge of operations. The Franklin Committee, after inspecting Ringwood in July 1768 at the request of the Royal Governor of New Jersey, painted an excellent picture of the efficiency and equipment with which the

Ringwood works was turning out iron at that time. The members reported that very little of the old works remained to be seen and nearly everything they saw was entirely new. They also compared the size of the blast furnace to the one at Charlotteburg and noted that it could make the same quantity of pig iron, about 20–25 tons per week. Although it was not in blast when the committee members visited Ringwood, they expressed their belief that it soon would be, as it was only awaiting the completion of the hearth which was being finished at the time of inspection.

Within 50 yards of the furnace, the committee observed, was a very good forge of three fires and two hammers, with a stamping mill which was cited as a particularly profitable piece of equipment. This mill was used to separate the iron from the cinder. Nearly 500 yards above the furnace was another very fine forge of four fires and two hammers with a sawmill near. A half mile below the furnace was a third forge with two fires and one hammer, while two miles further down the stream, was a fourth forge with the same equipment. All of these forges, the committee stated, were capable of making "250 tons of bar-iron single-handed, or 300 to 350 tons, double-handed, at each fire, of which there are, in all, eleven."

One of the greatest handicaps of the Ringwood works had been an inadequate water supply, particularly during seasons of drought. Work had to be suspended sometimes for several weeks, when the Ringwood River reached very low levels. Hasenclever had overcome this difficulty, however, by building a huge reservoir which greatly impressed the Franklin Committee, who described it in this way: "This reservoir is a pond, called Toxito Pond, is about three miles long and near a mile broad, it formerly emptied itself into the Ramapogh river, but by an immense dam of 860 feet long, and from 12 to 22 high, the natural outlet is stop'd up, and the water raised to such a height as to take its course with a head of 10 feet high into a new long canal which conducts it into the Ringwood River."

This report of the four-man committee also referred to the excellent quality of the ore in the mines that supplied Ringwood and to the abundant amount of wood available in the vicinity for fuel. It noted that since the Peters mine had been drained and another mine called the Good Hope on Wales Mountain had been opened on July 1, 1767, there was plenty of ore available and the quality of this ore was both "fine and tough." Access to the great number of trees, which it predicted would provide a never-failing source of charcoal, was provided by many roads that were to be found in and about Ringwood.

During the brief period that Jeston Humfray was in charge of the Ringwood works, little was accomplished. Hasenclever, of course, protested that Humfray had allowed the various parts of the iron-making machinery to deteriorate and had even altered the processing equipment resulting in inefficient production.

John Jacob Faesch, a highly-skilled German ironmaster, who had been brought over in 1764 by Hasenclever to supervise his iron-works in America, succeeded Humfray during the latter part of July in 1769. Under his direction production continued, although never on a scale that approached the full potential of the works as originally outlined and planned by Hasenclever. Faesch remained in sole charge of the American Company's works at Ringwood until the arrival of Robert Erskine on June 5, 1771. Erskine had been hired by the Company in England and upon reaching Ringwood he teamed up with Faesch to run the ironworks jointly until the summer of 1772. But friction between them caused Faesch to leave Ringwood and operate Mt. Hope, an ironworks he had purchased and at which he was building a furnace.[8] There Faesch made a name for himself as one of the Revolutionary ironmasters who made vital contributions to the Continental cause by supplying ordnance and other iron products Washington desperately needed for his army.

About a month after his arrival at Ringwood, Erskine wrote to Richard Willis, one of the partners in the American Company, about the conditions of the roads and of the works in general as he found them:

> The situation at RINGWOOD is tolerable, but nothing enchanting about it, the mansion house has been patched together at different times which makes it a very acquard [sic] piece of architecture. The furnace is in blast, but the bellows are old and not so good as at the other works, which were built under Mr. Faesch's direction, whereas Ringwood is a work of thirty years standing. There the water-wheel is exposed, but at Charlotteburg, etc., the wheel is under cover, and by

means of stoves may be kept in motion when the frosts are pretty severe.

This picture varies from what was reported by Governor Franklin's committee who, only three years earlier, had emphasized that "the present works" at Ringwood were "entirely new." Some see in Erskine's rather uncomplimentary remarks, his prejudice against Hasenclever as well as an alibi in case production did not come up to expectations. Erskine's letter written July 9, 1771, criticizes the location of the works in relation to shipping points for the iron:

> At Cortlandt there is enough water for a furnace, but not for a forge; in other respects, it is the only place for an iron works which Mr. Hasenclever in his hurry to provide a settlement for Mr. Faesch and his men, seems to have chosen with judgment, for the iron from Ringwood is carried: to Haverstraw, 25 miles–from Longpond to Hackensack, 27 miles–from Charlotteburg to Acquackanonck, 22 miles, land carriage–whereas the furnace at Cortlandt is within a mile of the great North River.[9]

It was probably the discouraging tone of Erskine's reports that prompted the American Company to consider disposing of these properties. An advertisement appearing in 1772 offered the three main works at Ringwood, Long Pond, and Charlotteburg for sale. Whether the partners of the American Company lacked the extra funds that Erskine felt were necessary to bring about a profitable operation, or whether they had become disillusioned over the proportionately small returns their investment yielded them, the decision was made to put the works up for sale. Undoubtedly it was with their approval that Erskine worded and inserted this advertisement in a New York paper:

> AMERICAN COMPANY IRON WORKS. NEW JERSEY
>
> Notice is hereby given that Reade and Yates in N.Y., or Robert Erskine Esq. on the premises, are impowered to receive Proposals either for the Sale or Lease of the Well Known Works of Charlottenburg, Long Pond, and Ringwood. Whoever, therefore, is inclined to treat for the Sale or Lease, either of Part or Whole, are desired to apply as above. Subjoined is a Description of the Works, taken by the desire of his Excellency Governor Franklin in the Year 1768.[10]

This advertisement evidently failed to draw any customers for the properties. The next year, Erskine advertised again in the New York papers trying a different approach for raising the much-needed funds.[11] He pointed out that he was authorized to receive all debts due the American Company and its ironworks. Lacking buyers for the mines, furnaces, and forges, he attempted to collect long-standing debts owed to the American Company in an effort to keep the works in operation. What measure of success he had with this venture is not apparent. At best, he probably received a mere trickle of money. But he managed to keep the Ringwood operations going, although not at full capacity, as a result of an agreement he worked out with Curson & Seton, a firm of New York merchants. Erskine agreed to supply Curson & Seton with all their requirements for pig and bar iron in return for cash. As surety for any overbalance of payments, he pledged "other stock of goods, moveable effects, horses and cattle" as well as additional pigs and bars of iron.

Although he had arrived in America only shortly before the Revolution, Erskine seemed to have caught the discontent of the Colonists under British rule. In his correspondence he sympathized with their rebellious spirit against the impositions of the mother country. For example, in a letter to his London employers written during October, 1774, Erskine pointed out that "the Oliverian spirit in New England is effectually roused and diffuses over the whole continent which, though it is now pent up within bounds, a few drops of blood let run could make it break out in torrents which 40,000 men could not stem much less the handful General Gage has, whose situation is far from agreeable." Six months before the first shot was fired at Lexington, he already foresaw and warned his employers that "the rulers at home have gone too far."

In another letter dated May 3, 1775, he described the events which were moving swiftly toward the conflict:

> The people, as I have said before in private letters, are sincerely in earnest everywhere. I have even been applied to for gunpowder by the principal people of the County of Bergen in the Jerseys, in which your Iron works are situated, where they, who till now hardly thought anything of the matter, are forming into regular disciplined bodies as fast as possible, which is the only business attended to at present anywhere. General Gage is shut up upon salt provisions in Boston, from whence it is allowed he could not stir 10 miles had he 10,000 men;

> for 20,000 men who now beyond doubt can fight, are entrenched without the town, and 30,000 more were sent home again as superfluous at present. . . . The present subject I have adopted from the general voice which held it necessary that all who corresponded with England should be explicit in declaring the situation of this country, which is beyond dispute indissolubly united against the British Ministry and their acts, to which the Americans will never subscribe but in characters of blood; nor since blood has been shed do I believe a hearty reconciliation can again take place unless blood seals the contract.

On August 17, 1775, three months after he had warned his countrymen in London of the dire consequences that would soon take place, Erskine was commissioned captain of a company of foot militia organized at the Ringwood works for defense against the British. This company, in Erskine's words, consisted of "forgemen, carpenters, blacksmiths and other hands, whose attendance is daily required." Realizing that this group of workmen could hardly be considered soldiers from the English viewpoint, he added, "I dare say, however, that there is not a man belonging to it but would willingly lend his aid in a case of extremity when every consideration must give way to the salvation of the country." [12]

Erskine's services to the Colonial cause were not limited solely to leading a company of raw militia. In 1776, he devised a type of *chevaux-de-frise* built of sharpened logs and ironwork that could be placed in the Hudson River as an obstruction for British warships. Since five British ships had passed the Continental batteries placed on Manhattan Island and at Paulus Hook, New Jersey, a short time previously, Erskine's letter outlining his invention and its advantages was welcomed as means of preventing British control of the strategic river. The *chevaux-de-frise* were placed in the Hudson channel at a point between Fort Washington on the Manhattan side, and Fort Lee on the New Jersey side. However, the ingenious plan to obstruct the river here was never carried out to successful completion. The river was deeper than anticipated and some of the wooden frames of the *chevaux-de-frise* were found floating with the tide.

Operations at the Ringwood works were wholly devoted, during the period of the Revolution, to making iron for military purposes of one type or another. Just how many employees Erskine had through most of this critical period is difficult to determine. Just prior to the outbreak of the war, in a letter to a friend in Scotland, he wrote: "I have 8 clerks, about as many overseers; forgemen, founders, colliers, wood cutters, carters and laborers to the amt. of 500 or 600. . . ." But desertions, runaways, and illnesses must have considerably reduced the number, particularly after hostilities started in 1776.

The difficulties he was having in trying to keep his woodcutters, while his forgemen were busy making component parts for another barrier is clearly indicated in a letter written February 8, 1777 to General George Clinton, Revolutionary governor of New York State and commander of the New York troops:

> I am sorry to inform you that the greater part of my woodcutters have gone off to the regulars; there have been frequently emissaries among them, the last I understand was an old gray headed man from about Hackensack. I am glad, however, to hear that about 23 of them have been taken this week in Bergen Woods, where they were Cutting fewel & hope it is true; meantime, I beg leave to observe that without wood I cannot again blow our Furnaces, that our stock of pigs will not last about half a year, and our Coals to about the middle of May; if, therefore, I cannot get a supply of hands we must stop; as the greatest part of the Iron works are stopt already, it is Certainly necessary for the service of the Continent to keep some agoing if possible.
>
> Our Forgemen are all at work making Irons for Chevaux de Frize, which we execute at the rate of 40 a day; the quantity, Capt. Machin, therefore, orderd will soon be finished, and any further orders for the same service shall be punctually executed.[13]

Erskine was frustrated in his attempts to get his workmen exempted from military duty. He repeatedly sought permission from General Washington for exemptions, pointing out the futility of trying to produce iron without manpower. This privilege had been granted to two of Erskine's neighbors in the iron industry, Colonel Ogden and Mr. Faesch. But Washington was apparently uncertain of the value of deferring all ironworkers from military service. According to Erskine, Washington was fearful of the results of blanket exemptions of this kind on military recruiting and thought that if he granted it to Ringwood, all private ironworks might demand the same favor. Yet Erskine obviously felt that Washington should be willing to make an exception in the case of Ringwood which was filling military orders at "great private loss, which the greatest price for years to come cannot make up." Furthermore, Erskine declared, "I know that the iron is dear now; it will be much higher if the works in question stop, as almost all others have done."[14]

Erskine's letters to Clinton throughout the year of 1777 give a keen insight into the affairs of Ringwood and its operations and the events and confusion of that time. They are filled with repeated references to payment for products turned out at Ringwood. As a result, the chaotic financial state of the Colonies can be clearly seen, as well as determined efforts of the military leaders to fortify every place against capture. The limited manpower available to the Colonies for filling so many jobs created a frantic and often wasteful misuse of abilities. Erskine's letter of March 14, 1777 to General Clinton is indicative of the wide scope of activities dealt with in this valuable correspondence:

> On my being disappointed the pleasure of seeing you at New Windsor, I left a letter and a Bill of the Chevaux de Frize Irons; concerning which I beg the favour of directions where to apply for the money, as I have immediate occasion for it. I beg leave likewise to observe, that on looking over my accounts relating to the Iron moved from Hackensack I find that I have received money from Mr. John Zabriskie for about 10 tons which he advanced, to reimburse himself by sales, which if he has not done, I shall have a Claim for 40 Ton of the Iron moved from New Bridge; the account of which I should be happy to have settled as soon as possible; If the Iron is to be used for Continental service, the sooner this is done the better; as Iron is rising, and it is reasonable I should receive the price it bears at the time of its being used and paid for.
>
> I received a letter last week, Dated Headquarters, from Major Clark aid-de-Camp to Genl. Green; wherein he notices there being a vacancy of the place of Cheef Engeneer in our army, kindly supposes me adequate to the department, says L'd Sterling meant to write on the subject, which I have not yet received, and desires an answer which he will lay before the General. I have accordingly wrote him that I cannot suppose myself qualified for such an office in many respects, particularly that part which relates to artillery, as I never saw a Bomb thrown in my life, nor a gun fired but at a Review or Birth day; but

> that branch, to which practical geometry and mechanics is necessary, I could undertake with some confidence, these studies having been both my business and pleasure. I further mentioned that my engagements here were such, that I was bound to abide by them both by the tyes of honour and gratitude; but that as I had little or no prospect of Carrying on the works, I might shortly be at leisure to devote myself to the Cause of Liberty in any way I could serve it, and mean time should be happy to render it all the assistance in my power . . .[15]

Although he declined the position offered him in this letter due to his lack of experience with cannon, Erskine's forthright statement about acceptable work was prophetic. He was to use his talents in "practical geometry and mechanics." On July 27, 1777, he was commissioned geographer and surveyor-general to the Continental Army, in which he made full use of his background. One of the most critical needs of the military forces was maps of the New York-New Jersey area in which the forces were deployed. In the little more than three-year period that remained of his life, Erskine demonstrated a remarkable facility for producing such accurate maps they are still used as models.

Erskine's primary occupation was still making iron. There could be no slackening in turning out badly-needed equipment and military supplies.[16] Yet he found—or made—time to draft two hundred of these war maps. This double burden undoubtedly contributed indirectly to his death. The strain and physical exertion required not only to supervise but to participate actively in both enterprises drained him of his strength and placed him under hardship in all kinds of weather. As a result he caught what was described as "a severe cold" in the early autumn of 1780 while on a mapping expedition. Erskine died only two weeks after his return to Ringwood from this trip. His grave there lies only a short distance from the former site of the forges, furnaces, and properties which he struggled to operate during the extremely difficult days of the Revolution. He had placed the Colonies' need ahead of any personal gain.

One of the very few source materials available that describes Ringwood and its inhabitants during the war years is the diary of Ebenezer Erskine, nephew of Robert Erskine.[17] After a perilous and rigorous trip from England, Ebenezer arrived at Ringwood on September 7, 1778, apparently not expected by his uncle. Yet the welcome he received and his impressions of the ironworks in the

Ramapos, as seen by the young man from across the Atlantic Ocean, are charmingly told:

> He [Robert Erskine] received me very kindly, also my companion, James Arthur that I had brought along with me. Next day he inquired into my views of coming to America when I gave him a genuine acct. of my reasons and manner of leaving Home. Upon which he desired me to give myself no trouble about what was past as I should find in him not only a Relation but a Friend, that I must stay with him till the present Disturbances were over which he hoped would be soon, that then he intended opening a Large Store in New York into which he would admit me a Partner, he likewise shewed me his Commission appointing him Surveyor & Geographer to the United States by which he was obliged to be most partly along with the Army & upon that Accnt. was happy he had got a Frd. to stay with Mrs. E. in his absence.
>
> At present he had no children, Mrs. E. having had the misfortune to have two miscarriages since she came to this Country but that now he believed she was in the way to have one soon. He also offered his Service to J. A. to get him a Berth in the Commissary Department, in ye meantime to make Ringwood his home till he was provided for.
>
> He told me that the Iron Works had mostly stopped as he now had only about 40 hands employed, that he used in former times to have between 3 & 400 in constant employ, that he still kept 2 Clarks here at Ringwood and one at Charlottenburgh and one at Long Pond, the first 16 miles & the other 4 miles from this. One of the Clarks at Ringwood was Robt. Monteith, an old acquaintance of mine at Edinr. He was Clark there with Andrew Sinclair & Hay and came to this Country about 6 years ago, had lived with Mr. E. ever since, the other Clarks name was Ambrose Gordon a this Country Lad but of Scots Extraction.

Among the details to be found about Erskine and his way of life in this diary of Ebenezer are the arrangements provided for safeguarding the surveyor-general during his many journeys during the war:

> Saturday 12th, Mr. E. set out for Camp along with a Servt. and 3 light horsemen who are appointed to attend him. Before he went away he gave J. A. fifty Continental Dollars, said he could repay him when he got into employ.

> October 10th I set out for Head Quarters to seek Mr. E. It was at Quakers Mile about 70 miles to ye Eastward of this upon the Borders of Connecticut. Arrived there the 12th, was introduced by Mr. E. to Genll. Washington who is a very affable good looking Man. Mr. E. gives me 100 Continental Dollars to keep my Pocket, stays there till ye 16th and returns to Ringwood ye 18th, had Light Horseman attending me going & coming.

Throughout the years of the Revolution, Ringwood offered a most appropriate target for British raiding parties. As a strategic center supplying arms and equipment to the Continental Army, its destruction undoubtedly would have been welcomed by the British command in New York City. There is a story, perhaps half-legendary, that a party of this kind tried to get through the pass in the Ramapos at Sidman's Bridge, on the outskirts of Suffern, New York, but was turned back by the small garrison stationed there by Washington throughout the war. However, Ringwood was raided on November 11, 1778 by a gang of thieves that included some of the members of the infamous Claudius Smith gang.

The following description of the event which took place while Robert Erskine was away from home is from the diary:

> On the 11th Novembr (1778) our House at Ringwood was Plundered by a party of Robbers. It was effected in the following Manner:
>
> The two Clarks Monteath & Gordon slept in a room off ye Counting House. J.A. & I slept in a room above ye Hall door in ye House.
>
> Between nine & ten in ye Evg. somebody knocked at ye Counting House door. Mr. Gordon asked who was there, was answered a small party of Continental Troops going to Morristown who wanted Lodging upon which he opened the door and immediately there rushed in two Men with Pistols with whom he had a short scuffle but 5 or 6 more coming in he was soon overcome. They threatened if he spoke a Word or made the least resistance they would instantly Blow his Brains out.
>
> They bound his Arms with a Rope and placed two Centerys over Monteath and him, they then came to the Hall door & knocked there. J.A. was gone to Bed, I was sitting reading, had heard nothing of what passed out of doors. I throwed ye Window up and asked who was there, they gave me for answer that they were 2 Gentn. who were benighted and wanted Quarters. I told them the Family were gone to Bed and they had better go to ye Counting House where perhaps the Clarks might accommodate them.

They said they had been there and Mr. Gordon was gone to Bed and had directed them into the house. I then went down Stairs to open the Door but being a little intimidated I went back and acquainted Mrs. E. (who was undressing). She desired I would let them in and shew them to the Room where Strangers usually slept. I then went down again and opened the Hall Door, when immediately Two Men presented a Pistol each of them in my Face.

I stammered a few Paces back and had almost fallen with the Surprize. They cryed don't stir or your a Dead Man, there was then five Men surrounded me with each a cocked Pistol presented at me. One of them says I seaze you in his Majesty's Name. I reply, what in the Name of God do you want, upon which he ordered me immediately to put the two Boys that slept in ye Garret and the Gentn. that slept with me with all ye Arms into their Custody, told me any resistance was vain, that he had been in ye Counting house and bound Gordon, that ye house was surrounded with 40 Men and 300 Indians in ye Mountains which he would bring down at ye Firing of a Gun, that Goshen was that night in ashes & if I made ye least hesitation of observing his Orders he would put me to Death on ye Spot.

I then told him his Orders should be complyed with as far as lay in my power only begged he would alarm pour Mrs. E. as little as possible. I then took them up Stairs, Called down ye two Boys (a Negro and Blacksmith Boy) whom they took into Custody. I next took them into my own Room where was J.A. lying trembling in Bed. They ordered him to lye still with which he readily complyed. They then seized upon two Muskets which was in ye Room and asked me if this was all ye Arms in the house. I told them it was all that I knew off, they said if they found any more I should suffer for them.

They then desired to be introduced to Mrs. E. I took them to her Room door (which she had bolted). She asked who was there, I told her it was some Gentn. who must see her but not to make herself uneasy as they had given their Word of Honour to be Civil to her.

She opened the Door but was much alarmed when she saw their presented Pistols. The Man that first addressed me told her he had the Command of ye Party and was come to search for Treasonable papers belonging to Mr. E. He also told her the Frightful Story of Goshen being in Ashes, of ye 40 men & 300 Indians, same time he told her that if she complyed with his Orders in making the Search she should not be hurt.

He then demanded the last Letters she had rec'd. from Mr. E. upon which she gave him one, he looked at it but told her she had rec'd. one from him since, that of ye 1st of ye Mth. She acknowledged she had and put her hand in her pocket and gave it to him which he without looking at folded up and put in his Pocket with the other one.

He then desired her to open a Trunk which stood behind her which she immediately did, Telling him there was nothing in it but some of her Cloaths. He turned it over a little & told her she might shut it again as he did not want any of her Cloaths (in that Trunk was 3000 Continental Dollars).

He then desired her to open a Desk which she did. He turned over a number of Papers and ye rest of ye Men began to pick out a Number of articles which was in the Desk viz. a Silver Watch, some Rings and earrings & c. belonging to Mrs. E. [They also took Erskine's contract as manager of the company's ironworks in New Jersey and New York. The loss of this document was a handicap later in settling the estate.]

He then told her he had certain information of a Chest of hard Money being here belonging to Coll Malcolm and if she did not produce it he would lay the house in Ashes. She assured him there was no such thing to her knowledge & if he would burn the house only begged he would let her and the other People out first.

He then demanded her Watch, to which she attempted to give evasive answers but upon their insisting she gave it up. They continued plundering the House of all the Plate, four Silver Watches, Mrs. E.'s Silver hilted Small Sword & c.

About one O'clock in ye Morning they order us all into the Parlour into which they also brought Monteith & Gordon. Gordons Arms was pinned with a Rope. They then began again to threaten Burning the house if the Chest of hard Money was not produced, but seeing these Threats had no Effect, they ordered Gordon, Monteath, Arthur and I down to a Cellar, locked the door upon us and took the key with them, they then broke the Locks of the Muskets which they did not take with them, they next ordered the Negro Cato to go to the Stable with them where they took out six horses on which they loaded their Plunder & set off threatening Mrs. E. that she should not let any of us out till six O'Clock, about two hours after they were gone Mr. Gordon takes a horse and sets out to alarm the Boards and other Neighbours about 3 miles off. Six in ye morning the Messrs. Board, Hoggin & c, six in all comes here armed, informs us that Mr. Gordon had proceeded to Pompton for a Party of Light Horse which comes here about eight when they proceeded down the Road we supposed they had taken, upon coming to Sidmans they inform the same party that robbed us had also robbed them and had left their house just at Day Break.

Mr. Gordon proceeds after them with the Party but gets no intelligence of them in ye afternoon and our six horses were found stragling on ye road four miles below Sufferns.

> Midday J. Arthur sets out for Camp to inform Mr. E. of our Misfortunes.
>
> The 15th J. A. & Mr. E. returns. Fir several weeks are employed in making Searches and have a number of People taken up on Suspicion but without any Effect. . . .

Several months after the robbery (on March 29, 1779), a confession by William Cole, a member of Claudius Smith's gang, mentioned the raid and named the men who participated in it. Claudius did not take part, for he was hiding out in Long Island at this time with a price on his head for killing Major Strong. However, some of his men did participate in the robbery. Cole stated that he left New York some time in the latter end of the last fall in the company of Thomas Ward, John Everett, Jacob Ackner, James Cowen, George alias Thomas Harding, David Badcock, James Twaddle, Martinus Lawson, Peter Lawson and John Mason, who was the leader of this particular group that night. Cole, "parted company with them in the Clove about a mile beyond Sidman's being something indisposed and remained in the house of Edward Roblin in the Clove, while the above named persons robbed Mr. Erskine and Mrs. Sidman in two robberies staged in the same evening." Cole also told how the above named George Harding had made a present of Mrs. Erskine's gold watch to David Matthews, Esq. the Mayor of New York City. He also stated that Mr. Erskine's rifle was given by Mason to Lord Cathart.

As the property of British owners, the Ringwood mines, works, and land were liable to confiscation under legislation passed by the Convention of New Jersey, formerly the Provincial Congress. Actually the resolution adopted by the patriot legislators on August 2, 1776 branded as traitors "all such persons within their respective bounds as have or shall have absconded from their homes and joined themselves to the enemies of this State." Since the London Company had never been residents of the state, a strict legal interpretation might have exempted them. There was no doubt, however, that they would have qualified as British citizens under the definition "enemies of this State." No mention can be found of confiscation of this property in either the records of the Commissioners for the Confiscation of Estates or in the *Lists of Confiscated Lands* in the County of Bergen. There are two probable reasons why: the ardent patriotism and valuable contributions of Erskine and the role of the ironworks itself as a critical source of essential materials and implements for the Continental cause. As long as Ringwood furnished items desperately needed to wage the struggle for independence, the absentee owners' political views were of little concern.

Following Erskine's death, a rather confused state of affairs existed in regard to the Ringwood properties. The American Company had not been in contact with Erskine since 1775. Curson & Seton had been dropped as agents after a dispute with the proprietors in 1774. Erskine himself took over the agency and had been in complete control. With the outbreak of war, the English ownership had virtually been wiped off the books, and the question had become one of responsibility and actual possession. The Ebenezer Erskine papers at the New Jersey Historical Society reveal that "the difficulty of procuring hands and the necessary supplies for them had nearly closed the business of the works before Mr. Erskine's death: that event finally stopped them and with very little disadvantage to the company. There are now no more hands employed about the works than are necessary to collect the stock and property together and to take care of it and a clerk to close the books."

Within days after Erskine's death, Samuel Ogden, a member of the prominent family which had been associated with the Ringwood properties as early as 1740, received a letter from William Seton, one of the partners in Curson & Seton, which he delivered in person to Mrs. Erskine at Ringwood. In this letter, dated October 16, 1780, Ogden was empowered by Seton to "take the whole [Ringwood works and properties] into your charge and care, 'till such times as you can be furnished with the proper and necessary powers." Seton, acting solely on his own, "having just heard of Mr. Erskine's death," stated that "Mr. Richard Curson and myself are the last Agents here from the Proprietors in England and the only persons that have the power to take charge of their affairs in this country." Seton also asked Ogden, a former justice of the Provincial Supreme Court before the Revolution as well as a prominent ironworks owner, to notify Richard Curson, Seton's partner, "who lives at Baltimore in Maryland." This seems like a rather curious request. Evidently Seton and Curson were not on close friendly terms or, perhaps, Seton believed that no time should be lost in trying to get control of the valuable Ringwood estate.

Mrs. Erskine now must have wished she had her husband's stolen contract with the company. However, the attempt of Curson & Seton to take over the Ringwood property failed. At least there is no evidence to show that the plan was carried out. It is possible that the Patriot legislature and Governor William Livingston thwarted the plans of the New York firm. Mrs. Erskine not only had physical possession, but also seems to have had a thoroughly loyal group of employees and friends who resisted any attempts to disposess her. In addition, her friendship with General Washington through her husband's close association with him, undoubtedly was a decisive factor.

Mrs. Erskine's marriage in September of 1781 to Robert Lettis Hooper, Jr., one of the owners of the famous Durham ironworks and a Deputy Quartermaster General in the American Army, served to protect her from any further danger. Eight months later, on May 20, 1782, the New Jersey Legislature passed an act, possibly through the influence of Hooper, giving him and his wife, powers of agency to take charge of and manage the estate of the American Company. This act did require the Hoopers to make periodic accountings to the Legislature, but it barred forever the rather unlikely claim advanced by Curson & Seton to possession of the famous ironworks in the Ramapo Mountains.

Just what steps Hooper took to rehabilitate the forges and furnaces of Ringwood after he was given the right to operate them by the New Jersey Legislature cannot be determined by examining available materials covering this period. He was probably either unsuccessful in his attempts or he simply decided that it would be economically impractical to start producing iron again. At any rate, in an advertisement in the *New Jersey Journal* of April 23, 1783, signed by Hooper, there are indications that Ringwood would have no further need for most of its equipment. Among the items offered for sale in this notice were such varying objects as featherbeds, furniture, a wide range of mining equipment and tools, and livestock that included horses and milch cows. The household articles were advertised for public sale, while the "refined bar-iron, carpenter's and joiner's tools, several pair of furnace and forge bellows, in good order, large assortment of furnace and forge tools, vises, hammers" and all other iron-manufacturing or industrial pieces were earmarked for private sale only. In a final paragraph, Hooper appealed to "ALL THOSE INDEBTED TO THE AMERICAN RINGWOOD COMPANY by bond, note or otherwise" to settle their debts "at Ringwood on the first day of May next, or they will be proceeded against as the law directs. . . . Those who have any demands against said Company" were asked to present their accounts at the same time for settlement.

Whatever the success of the latter appeal or the outcome of the public and private sales, the Hoopers moved to their estate "Belleville" at the falls of the Delaware River near Trenton, New Jersey, shortly after the peace treaty had been signed between the newly-created United States of America and Great Britain in April of 1783. Ebenezer Erskine also left Ringwood at approximately the same time and moved to the Trenton area. The once mighty iron empire which Erskine had labored to expand to its full potential and convert to a thriving industrial enterprise became deserted. The mines, the hammers, the furnaces, and raceways lay idle and neglected. The numerous charcoal burners, blacksmiths, carpenters, colliers, and workmen of one sort or another drifted away into the sloping forested acres of the surrounding Ramapo wilderness. Once again—and not for the last time in its long and colorful history—Ringwood was abandoned.

The fifteen-year period following the departure of the Hoopers and Ebenezer Erskine from Ringwood has always been considered a missing link in the background of this famous ironworks. Albert Heusser, the biographer of Robert Erskine, has termed it "a puzzling hiatus." Yet at this time Hector St. John de Crèvecoeur, a famous French traveler and writer, visited the Ringwood, Sterling, and Charlotteburg properties. At least his descriptive accounts carry the date 1789–90, although there are discrepancies in the writing that tend to disprove his accuracy. Crèvecoeur, whose earlier book, *Letters From An American Farmer,* became a classic because of its graphic descriptions of Colonial life in America, wrote an intimate, if somewhat suspect, tale of Ringwood and its manager just before the turn of the century in his *Voyage dans la Haute Pensylvanie et dans l'Etat de New-York.*

Crèvecoeur quotes a "Mr. Erskine" at the time he and his friend "Mr. Herman" visited Ringwood. Just two days earlier Crèvecoeur had been at the Sterling ironworks where the proprietor, Peter Townsend, told him he had just finished making some new lightweight plows for General Washington who "is filling with his so distinguished talents the Presidency of the Union," and at the same time "looks after the immense tillage and operation of his estate and farm" at Mount Vernon. As Robert Erskine died in 1780 and Washington did not become President until 1789, Crèvecoeur could not have spoken with Erskine at the time he reported. And it makes one wonder when—or whether—he ever did. Since it has also been established that Ebenezer Erskine had moved from the vicinity long ago he could not have been the "Mr. Erskine" mentioned either.

"Last year," wrote Crèvecoeur, "Mr. Erskine told us that he sold five hundred tons of iron in bars, two hundred of steel without counting the cast iron." No records have been found so there is, of course, no way of checking on the amount of bar iron, cast iron or steel that was made in the years 1789–90. But the very lack of statistics and information regarding Ringwood and its production, particularly in this period, suggests that these figures may have been figments of an overactive imagination.

Heretofore, the only document brought to light that reveals any information on the Ringwood area during this 15-year gap is a contract uncovered by Albert Heusser. On June 25, 1795, according to Heusser, Sir George Jackson and Robert Mure (or Muir), "trustees of the American Iron Company," through their agent, Phineas Bond, contracted with James Old of Pool Forge, Lancaster County, Pennsylvania, for the sale of the Ringwood estate (including Long Pond but not Charlotteburg)—about 12,000 acres. The sale price is given as 9,000 pounds sterling to be paid in four installments during the years 1796–99. Old was apparently unable to buy the property alone and tried to interest some Pennsylvania ironmasters in portions of the property, but without success.

Heusser says that James Old shifted the burden of the transaction to his son, John. Meanwhile, a deed dated February 6, 1796 was filed by "William M. Bell, Esq., High Sheriff of Bergen County to John Travis, Merchant of Philadelphia," in which Bell, as the result of "a writ of the State of New Jersey issuing out of the Supreme Court of Judicature held at Trenton at suit of John Jacob Faesch, Esq. [against Sir George Jackson and Robert Mure, seized] all lands in the County of Bergen generally known by the name of lands and property of the American Iron Company" and sold them, presumably by auction, to Travis. The tracts described in this deed were at Ringwood, Long Pond, and Charlotteburg and total approximately 18,000 acres.

For some unknown reason, Heusser apparently failed to discover the deed of February 6, 1796, for if he had, he would have noticed that it contains a reference to an earlier deed, which would have helped him fill another small gap of the missing period. This earlier deed, dated August 23, 1793, two years prior to the uncompleted sale of Ringwood to James Old by Jackson and Mure, was drawn up between Thomas Lloyd and Daniel Sill; and Sir George Jackson and Robert Mure, for all those tracts situated at Ringwood, Long Pond, and Charlotteburg contained in release made by Peter Hasenclever on February 26, 1767. Each of the tracts was subdivided, with the principal acreage in each 6018.88; 5937.38; and 5319.13 acres respectively. Presumably this earlier deed made by the same trustees of "the American Iron Company" has been overlooked or unmentioned because Lloyd and Sill either did not succeed in obtaining a clear and unblemished title to the properties, or for other reasons, perhaps financial, their possession of the mines, forges, furnaces, and acreage was never actually confirmed. The sale to James Old, however, is one of a series of connecting links in the history of the Ringwood Company that can be clearly traced down to the present time.

Despite the fact that Jackson and Mure had a contract for the sale of the Ringwood real estate with James Old, made on June 25, 1795, Sheriff Bell, as pointed out before, sold the property out from under these trustees to John Travis, thus leaving Old apparently without a legal claim. Old, however, was not to be denied. Almost two years after Travis purchased the lands and property of the American Iron Company, he sold the Ringwood and Long Pond tracts, giving a corroborative deed dated January 1, 1798 to "John Old [son of James] of Ringwood, Iron Master." Old was not to be fooled again in his second purchase for the deed cites the circumstances under which Sheriff Bell seized the properties, held an auction, and sold them to Travis. But no mention is made, nor can any further evidence be found regarding the sale to Lloyd and Sill and the sale to Old was never completed because he went into debt and could not pay the purchase price.

In the years immediately following the first inauguration of George Washington, impetus was given to commercial and industrial growth in the young nation by its rapidly increasing needs. One of the better known plans for taking advantage of this expansion was Alexander Hamilton's Society for Establishing Useful Manufactures. It had selected as the site the Passaic Falls, a short distance south of Ringwood, but there is no document in existence to show that any enthusiasm for the plan was aroused in the Ringwood area. If John Old had rehabilitated his mining and processing plant in the Ramapos, he might have been able to capitalize on the rising boom. An advertisement placed in various issues of the *New York Herald,* from November 5, 1803 to April 28, 1804, offering the Ringwood and Long Pond ironworks for sale is proof that Old was unable to profit from his properties. The advertisement pointed out the numerous mines on the land and the two tracts totaling 12,800 acres with cleared land amounting to 600 acres and good meadow land 250 acres. At Ringwood "a large Iron Furnace in complete order . . . has cast four hundred and fifty tons of cast iron a year. One Iron Forge built last year, in perfect repair . . . has three fires and is equal in point of situation and repair to any forge in New Jersey."

It is interesting to note that Old also mentioned in his advertisement several sites for waterworks on the stream running through the Ringwood tract, "some of which were once occupied by the American Company [the former proprietors of the estate] as seats for forges." It was pointed out that this location had the distinct advantage of being located within 25 miles of three different landings; "Acquacknonk, on Passaick river; Demarest's Landing on Hackinsack river, and Haverstraw, on the Hudson River." The second landing mentioned was in what is now River Edge.

All in all he painted a favorable picture in his advertisement, even describing the manor house in some detail. He must have been somewhat disappointed, therefore, when it failed to bring the desired result. Once again, the potentially rich and productive investment literally went begging. Less than nine months later, Old lost the properties through foreclosure proceedings inaugurated by Casparus Bogert, Sheriff of Bergen County in favor of James Lyle, to whom Old owed money. The date of this deed transferring the ironworks to Lyle was August 28, 1804. The new owner held the property only until March 18, 1807, when a local resident, Martin J. Ryerson of Pompton, New Jersey,[18] acting through his attorney, David B. Ogden of New York, bought one tract of land on the Ringwood River containing 6,838 acres and another on the Long Pond River of 5,975 acres.

With the sale of the estate to Ryerson, Ringwood at last came into strong and capable hands. Ryerson, an experienced ironmaster, who was already operating the Pompton ironworks, opened up a number of new mines in the Ringwood section, increasing the flow of ore. Through efficient planning and policies that included utilizing all resources to their utmost, he again made Ringwood into a going-business operation.

For years, Ryerson working Ringwood turned out iron of an excellent quality. An idea of the scale of operations carried on there

---

TO BE SOLD, that valuable estate known by the name of RINGWOOD IRON-WORKS, situate in the county of Bergen, and state of New-Jersey.

This estate contains about six thousand two hundred acres of land, of which there is upwards of six hundred acres cleared; two hundred and fifty acres of good meadow, which is now regularly mowed; besides a large quantity of land capable, with a little expence, of being converted into excellent arable land and meadow. There is on the premises a large Iron Furnace in complete order, which has cast four hundred and fifty tons of cast iron a year. One Iron Forge built last year, in perfect repair, which has three fires, and is equal in point of situation and repair to any forge in New-Jersey. On the stream of water running through this tract there are several other scites for water-works, some of which were once occupied by the American Company (the former proprietors of this estate) as seats for forges and a grist mill; others have never been occupied at all. The wood on this tract is excellent, and in inexhaustible quantities, within one mile of the furnace; besides coal-houses, workmen's-houses, and all other buildings necessary as appurtenances to the iron-works.

There is on this estate an elegant Mansion House, 92 feet front, and about 30 feet deep, on which there has lately been expended a large sum of money, and is now in perfect repair. The Wood also on this estate is sufficient for every purpose which can be wanted about the works—In short, when it is considered that these works are within 25 miles of three different Landings; Acquacknonk, on Passaick river; Demarest's Landing, on Hackinsack river, and Haverstraw, on the Hudson River—it is believed that no estate now offered for sale affords so fair a prospect of advantage to a purchaser as this.

Also, the tract of Land known by the name of the Long Pond, adjoining the former, containing about Six Thousand Six Hundred Acres. This tract was also formerly the property of the American Company, and they had on the premises a large Iron Furnace, a Forge, and other works. The Iron Mines on this estate are numerous and good; the Scites for Water works are equal, if not superior, to any others in New Jersey; the wood in great abundance; and the estate one of the best objects of speculation now at market. These two estates will be sold either separate or together, as may best suit the purchaser. If, however, they are sold separate, the Long Pond tract will be sold first. For terms apply to the subscriber at the Ringwood works. The payment will be made easy to the purchaser.

JOHN OLD.

N. B. If the above property is not sold at private sale before the 16th day of April next, it will on that day be offered at Public Sale, at Ringwood Furnace. At the same time and place will be sold a variety of Household Furniture, a number of Team and Hackney Horses, Oxen, Milch Cows, Waggons, Carts, two complete setts of Smiths, and one of Carpenters Tools, &c. &c. The terms of sale will be made known by applying to the Subscriber at or before that time.

J. OLD.

Ringwood, March 15. March 21 to Ap. 16

The *New-York Herald*, April 4, 1804.

can be seen in an inventory taken shortly after Martin Ryerson's death, August 19, 1839, when the property went to his son Jacob:

1.10.0.0	bar iron in new store	@	$70.00 per ton	$105.00
1	forge hammer face (65 lbs.)	@	.05 per lb.	3.25
	scale beam & weights in grist mill			15.00
	small scale beam and iron forks			5.00
6	kegs powder	@	3.25 per keg	19.50
	anvil pattern			10.00
	pine lumber and ore box			8.00
4.11.0.0	bar iron at Ringwood Forge	@	75.00 per ton	341.25
16,390	bushels of coal at Ringwood Forge	@	.05 per bu.	819.50
	tongs, scales & implements at Ringwood Forge			35.00
4½	tons inferior hay at Ringwood Meadow	@	6.00 per ton	27.00
200	tons Mule ore at the mine	@	1.00 per ton	200.00
50	tons Mule ore in the kiln, now burning	@	1.00 per ton	50.00
20	boyers at the Mule Mine			7.50
120	tons Peters ore at the mine on the bank	@	1.00 per ton	120.00

But this period of prosperity did not last long after Ryerson's death. His sons who inherited the properties were not as successful as their father and were soon in debt. By an act of the New Jersey Legislature passed March 4, 1842, The Ryerson Iron Company was incorporated. Five months later, in August of the same year, part of the Ringwood land was listed for Sheriff's sale. Soon afterwards all of it was advertised for sale, but buyers were evidently scarce. For a ten-year period ending in 1852, the sheriff of Bergen County continually made plans to auction off the properties to the highest bidder only to postpone the sale each time. After a protracted length of time, all of Ringwood including the land was sold on September 1, 1853 for $100,000 to Peter Cooper, the noted industrialist and financier of New York.

Cooper's negotiations for the sale were carried on by Abram S. Hewitt, a rising figure in the industrial world and the business partner of Cooper's son, Edward. At that time, Hewitt was both secretary and business manager of the Trenton Iron Company. A shrewd opportunist, he not only was thoroughly familiar with the iron business but was also aware of the profitable potential of the Ringwood works.

The somewhat complicated transactions involving the sale of the properties are explained in two letters by Abram Hewitt to David Ryerson, written June 17 and June 24, 1853: [19]

> The Ringwood property was sold yesterday. I bought it for Mr. Peter Cooper for the sum of $67,000, being an amount sufficient to pay all the liens off. I expect to arrange with Mr. Jacob Ryerson next week for the property. I paid 10% down and arranged for the deed to be delivered 1st Sept. when the balance of money is to be paid. This removes the necessity of any writing between you and Mr. Cooper, as on 1st September the Sheriff will be ready to pay off the mortgages.

Some question must have arisen in Hewitt's mind or, he may have been queried by David Ryerson regarding more exact details of the transaction as outlined in this letter. Seven days later he again wrote Ryerson:

> I see that my last letter was not quite explicit enough. Mr. Cooper bought the property for $67,000 and gave Martin Ryerson a bond that he could reconvey the property at any time before 1st September, on being repaid his advance with interest. If this is not done, Mr. Cooper will be compelled to complete the purchase on 1st September. The effect is to give "Martin" a chance to make a good sale between now and then. Whether we shall buy or not, I do not know.

Ryerson's sons were obviously so deeply in debt that Cooper and Hewitt felt sympathetic enough to allow them the chance to sell the property for a larger sum of money before exercising their option to buy. But the $67,000 was only a partial payment of the final sum. Hewitt wrote to Robert J. Walker, a stockholder in the Trenton Iron Company: [20]

> After a protracted negotiation, Martin Ryerson has offered to sell the Ringwood property to the Trenton Iron Company for one hundred thousand dollars. We have examined the property carefully. We are satisfied the ore is really inexhaustible and of good quality. We entertain little doubt that the Penna. Coal Company will construct a road direct from the Pittston coalfield to Hoboken, passing through or near this tract, by which its value will be doubled. That Company has just authorized us to have surveys made at their expense and that they will proceed with the enterprise just as soon as they can secure from the State of New Jersey the necessary legislation.[21]

Hewitt wasn't altogether satisfied however: he thought he should explore the possibilities of more economical and improved methods of transportation. In a letter to Ephraim Marsh, president of the Morris Canal Company, July 19, 1853, he presented a leading question:

> The Trenton Iron Company is about to become the owner of the Ringwood tract, which they plan to work on a very extensive scale. Two courses are open:
>
> 1. Build a railroad to Pompton and transfer the ores thence by the feeder and Morris Canal to their furnaces at Phillipsburg.[22]
> 2. To erect furnaces at Ringwood and obtain a supply of fuel by the Erie Railroad, which can be had next year on satisfactory terms.
>
> At present the condition of the Pompton Feeder and the rate of tolls would preclude the possibility of adopting the first course, which we all prefer. I am instructed therefore to ask on behalf of the Trenton Iron Company:
>
> 1. Whether you will make the Pompton Feeder navigable for the same class of boats as the rest of the canal and keep the same in navigable order etc.?
> 2. What rate of toll on ore brought to the Pompton Feeder from the Ringwood tract?
>
> You will readily understand the necessity for these questions, for to construct a Railroad will cost one hundred thousand dollars. The quantity of ore annually brought down should be from twenty-five to fifty thousand tons to justify this expenditure. It is a simple question between you and the Erie Railroad. Nature favors the latter.

Meanwhile Hewitt sent P. R. George, head miner of the Trenton Iron Company, to inspect the Ringwood area carefully and report on its possible future value. According to George, the properties were reasonably priced and by working all resources to their greatest extent, the estate could soon pay for itself. Following this report, Hewitt wrote what must have been extremely good news to Martin Ryerson at Sloatsburg, New York, July 25, 1853: "The Trenton Iron Company has decided to purchase the Ringwood property, including all the small tracts, which have been under discussion at various times, for the sum of $100,000, payable $60,000 in cash and $40,000 in bonds of the Company, bearing 6% interest etc. . . . all to be paid off by six years time."

In addition, Hewitt stated, the purchase must date from September of the same year; the steam engine on the property would have to be included in the sale; and possession of the mines was to be taken at once to prepare for production. Later he wrote to Thomas Wallace at the Pompton ironworks August 19, 1853: "We have had a conference with Mr. George and he thinks there will be no difficulty in furnishing you with 3,000 to 5,000 tons of Ringwood ore between now and spring. . . . We are willing to sell at $3.00 per ton cash, on the mine bank. We have determined to manage Ringwood on a cash principle and make it take care of itself. If it cannot do that, it will be a poor purchase."

The transfer of the Ringwood property was made as scheduled on September 1, 1853. The amount of ore that had been mined on this old and valuable property prior to this date is estimated at from 300,000 to 500,000 tons, but there were still tremendous quantities unmined. For the first few years the mines as well as the forges were worked to some extent by George and his men but the furnace at Ringwood—in ruins for over thirty years—could not be used. These limited operations helped to earn interest payable on the tract's price. At this time the cost of putting up blast furnaces or building railroad spurs to the works was too high to be economically feasible. Increasing the volume of the Ringwood resources would have to await more auspicious times while the main problem to be solved—transportation—was being considered.

Hewitt was unable to get a favorable rate from the Morris Canal Company nor was the Ringwood Valley Railroad, as it was tentatively called, ever built. A route for this line had been mapped out from Ringwood to Pompton. Some of the letters which Abram Hewitt wrote to the Morris Canal Company reveal the tactics he used in an attempt to lure them into accepting his plan. In one, he pointed out that "a fortunate concurrence of circumstances puts it in your power to secure a large business which you had no right to expect." He then proceeded to tell them that the "resolution" they drew up did not conform with the thinking of the Trenton Iron Company and he presented a contract he had drawn up for them to confirm.

By 1856 he still had not convinced the canal company that they should lower rates, but Hewitt did not give up easily. In still another letter, dated June 18, 1856, he baited them again with prospects of gain: "It looks like we will build a Railroad to join the one at Sloatsburg. We are now making 1200 tons of blooms alone per annum, all of which will go to the furnaces (at Phillipsburg), then to Trenton. Mr. Brown of Waywayanda Furnace sends his by the Erie Railroad, but will change if you lower your rates . . . surveys have been ordered made to the Erie."

The reply to this letter evidently was not whole-hearted agreement with Hewitt's proposals for Hewitt answered it with the following:

> If new conditions are perpetually introduced we will never be able to come to an understanding . . . we propose to do the following:
>
> A. To begin the Railroad, but we can not agree to finish it.
> B. To cart 5000 tons per annum and not 10,000 tons until you are refunded in tolls the whole cost of the feeders operation.
> C. A guarantee that this cost shall be refunded in five years out of the tolls.
>
> We are prepared to enter into an agreement . . . to avoid all further difficulty, we decline to agree to bring 10,000 tons per annum to the Canal, although we expect to take that quantity . . .

The conditions prevailing at Ringwood in 1857 when Mrs. Hewitt moved there were described to her son, Edward Ringwood Hewitt.[23] She said that the lands were littered with the dross and other remains of an iron business that had been in operation for over a century. Slag heaps and cinder piles were all around, as well as assorted buildings in various states of disrepair. According to her son, she was greatly discouraged at times because it took more than twenty-five years to restore the properties to some semblance of a private estate, rather than a place of business.

None of the ponds in front of the manor house were in existence when the Hewitts moved to Ringwood. The water from the furnace waterwheel ran into a brook which flowed through the meadows to the main stream—the Ringwood River. The large pond, now in front of the house, was created about 1895. The main road to the mines from the furnace passed the house below the sloping bank. The first building along this dirt road was the old country store, a center of attraction in the community for more than a hundred years. In 1857, where this road turned south toward the family burying ground, two houses stood near two large oak trees. In the previous century this site was occupied by the magazine where powder for the mines was stored and supplies for the Continental Army kept. Along the road, where it passed the graveyard, a row of workmen's wooden houses and some log huts stood, but these were cleared away by Mrs. Hewitt's orders after she arrived at Ringwood.

Abram Hewitt had Ringwood thriving once more. His success, judging from a letter which he wrote April 24, 1857 to Mark Healey of Boston, must have encouraged him to consider the possibilities of greater expansion and growth:

> We ought to have two more blast furnaces and the Ringwood ores made accessible, and we could then be sure of making 40,000 tons of pig iron per annum, which could yield $250,000 a year even if the other property yielded nothing. We could do this by spending $200,000. It is a pity not to do it, but the stockholders must have their dividends. If we could have them take stocks or bonds in lieu of cash, we should be able to put up these furnaces and build the Ringwood Railroad. We have a very valuable property, but it is not yet developed . . . the responsibility is a heavy one.

Hewitt was still concerned with the most economical way to transport ore from Ringwood. All attempts to have the Morris Canal Company see things his way were in vain. Now he was courting the Erie Railroad in a never-ending search for opening up an outlet through which the Ringwood products could be funneled. His estimate of the value of a rail link between the mines and the Erie's main line at Sloatsburg is given in a letter to J. C. Kent "Agent" at Easton, Pennsylvania:

> The Erie Railroad will agree to take the ore from Ringwood to Jersey City for 75¢ per ton. We can get them freighted by canal for 75¢ freight and 50¢ toll, making $2. per ton from Ringwood to the furnaces, as soon as we can get a railroad from the mine to the Erie. The cost of the Railroad will probably be $100,000, but it will give an equivalent value to the property aside from the ores. We must get this road built.

At the same time Hewitt was constantly seeking new and improved ways of making iron that would command higher prices and bring about increased production and sales. On July 30, 1857, writing to Mark Healey again, he revealed some successful attempts at improving the iron that might yield benefits: "The experiment in making wrought iron direct from the ore with anthracite coal is very promising and I believe it will succeed . . . in this case Ringwood will have extraordinary value, as we could make blooms at $25. per ton or less and save at least $15. per ton. The result will be that we can make plate iron as cheap as in England and iron ships can then be built here as cheaply as there. . . ."

All the optimistic forecasts and hopes which Hewitt held for the future of Ringwood suddenly vanished in 1860. Despite the efforts made to market Ringwood iron, the Cooper, Hewitt & Company enterprise found itself with few customers. On April 24, 1861, Hewitt wrote: "It is painfully evident that it will be extremely difficult to sell the products of our works. I think therefore that it will be necessary to stop mining at all the mines except the Allen." He proceeded to instruct his superintendent at Ringwood, P. R. George, to "arrange for the cessation of all work, except the Allen Mine on May 1st . . . make no arrangements to work the Hibernia Mine till I see you. The financial difficulties are so great here that it will be necessary to avoid the expenditure of a single dollar that can be saved."

On the verge of disaster, Ringwood was saved by the Civil War. The famous old iron center revived when an order arrived during the latter part of 1861 for gun carriages to be used to support mortars. The order emphasized speed. There was a special knack involved in building these mortar beds. The tremendous recoil that took place when a mortar was fired imposed a severe strain on the earthwork, concrete block, or base on which it was placed, because of the extreme oblique angle assumed by the barrel of the gun. In fact, this recoil was so intense that it was capable of causing serious damage to the base: in the case of naval warships where the mortar was fired from the deck, the recoil was powerful enough to shatter even the steel plates upon which it rested. Hewitt, with the assistance of his staff of workmen, was able to build a mortar bed that could absorb a portion of this kickback thus reducing the danger to the firing base. Completing this order in record time, he gained a solid reputation with the ordnance department not only for speed but also for the highest quality workmanship and product engineering.

In 1896 Hewitt wrote a letter, recalling a remarkable telegram. It had been sent when Grant was about to attack two Confederate forts. No doubt the reason for appealing to Hewitt was the reputation he had made for rapid completion of ordnance requests:

> I am told that you can do things which other men declare to be impossible. General Grant is at Cairo, ready to start on his movement to capture Fort Henry and Fort Donelson. He has the necessary troops and equipment, including thirty mortars, but the mortar-beds are lacking. The Chief of Ordnance informs me that nine months will be required to build the mortar-beds, which must be very heavy in order to carry 13-inch mortars now used for the first time. I appeal to you to have these mortar-beds built within 30 days, because otherwise the waters will fall and the expedition cannot proceed. Telegraph what you can do.
>
> A. Lincoln

Impressing its importance on his employees, Hewitt proceeded with this essential task. In exactly three weeks (or nine days before the limit set by the President's telegram) Hewitt accomplished the seemingly impossible. He even called on mills other than his own to complete the job. When the time came to load the mortar beds on railway cars for shipment, Hewitt was determined not to have the railroads lose any of the valuable time he had gained for Lincoln and Grant. On each car was painted in large letters: "U. S. Grant, Cairo. Not to be switched under penalty of death." [24]

Hewitt took great pride in this accomplishment. An excerpt from a letter written to P. H. Watson, the Assistant Secretary of War, is evidence: "the most memorable mechanical achievement so far as time is concerned, that we have ever witnessed." There was still however another important matter connected with the job that required some attention. The entire cost of this undertaking—$21,000—had been borne by Cooper, Hewitt & Company, even though the company was in none too favorable financial condition. Hewitt, in a letter to General J. W. Ripley of the Ordnance Bureau, declined "all compensation or profit in this business, being glad that our knowledge and position can in any way be turned to account in the present crisis of our national existence." While his patriotic motives kept him from seeking a profit on the speedy military contract, Hewitt was eager to be paid for the cost. Not hearing from the government, and after a second letter to General Ripley, Hewitt decided to make a visit to the national capital to lay his case before the President.

Merely by having his calling card sent in, Hewitt was able to see Lincoln ahead of others waiting. According to Hewitt's own account of this occasion, Lincoln expressed surprise when he learned that Hewitt had not been reimbursed for the job and immediately sent for Edwin M. Stanton, his Secretary of War. Hewitt quotes Lincoln as saying to Stanton: "Do you suppose that if I should write on that bill 'Pay this bill now', the Treasury would make settlement?" Stanton shrugged his shoulders, Hewitt writes. Having sent for the bill Lincoln proceeded to put on the bottom of it "Pay this bill now. A. Lincoln." Then he said to Stanton: "Now, Mr. Stanton, I want you to do me a service. I am going to trouble you to go to the Treasury Department with Mr. Hewitt and see that this bill is sent through the proper channels for immediate payment." [25] Hewitt wrote on March 24 that following this interview he received a draft on the federal government for the full sum of $21,000. The difficulty in obtaining this money did not deter him from accepting and completing similar contracts. Among the forms of ordnance forged for the Union by Cooper, Hewitt & Company, were gun-barrel iron, armor plate for gunboats, and for the famous ironclad *Monitor*. One of the proudest possessions of Abram Hewitt was a document written by Stanton acknowledging the loyal and patriotic services rendered to the Union during the Civil War and thanking Cooper, Hewitt & Company. Hewitt had it framed and hung it in the manor house at Ringwood.

Under wartime production, the Ringwood properties had turned out essential arms and equipment without particularly worrying about the means of transporting it from Ringwood to its destination. In 1864, however, Hewitt's mind turned once again to his major problem of providing an economical and simple means of shipping from Ringwood. He wrote to the Morris Canal Company again: "The Sterling Property has been sold to a very strong Philadelphia Company. Reduce your rates and get both our business . . . [as much as] 100,000 tons could be sent. You will lose a great deal of business by the Morris and Essex Railroad. Here is the substitute. . . ."

But once again Hewitt's inducements to the Morris Canal Company came to naught.

During the Civil War period, mining was done at Hibernia, the Allen mine, and Ringwood where several mines were worked. Although Cooper, Hewitt & Company had their headquarters at Ringwood proper, the only active blast furnaces were at the Long Pond works, near what is now Hewitt, New Jersey.[26] The name Long Pond came from the nearby body of water (now Greenwood Lake). Hewitt apparently considered the Long Pond works to be a part of the Ringwood ironworks, since they were so close to each other. He often spoke of the furnaces at Ringwood in his correspondence, whereas they actually were at the Long Pond site, although in the general Ringwood area.

There had been a post-war decline of business at Ringwood which lasted until 1868. On April 15 of that year, Hewitt bought out the Ringwood interest owned by E. R. Miller, who had become a partner on November 1, 1861. Anticipating increased business, Hewitt ordered the furnaces at Long Pond, which had been standing idle, started up again and ore began to flow from the mines once more. This expectation of increased sales was short-lived; by 1870 the iron market had turned sour and once again he was ready to suspend operations. The Ringwood enterprises had a $12,000 loss on its books for its 1869 operations and a total loss of $25,000 for both 1869 and 1870.

But Hewitt's indomitable spirit and rugged business sense did not allow his faith in the quality of the Ramapo Mountain ore to waver. On July 8, 1871, trying another approach to the long-standing transportation problem, he turned to John Taylor, freight agent of the Lehigh Valley Railroad, with a proposition:

> We own the great Magnetic mines at Ringwood, from which the Midland Railroad is now being built to Mead's Basin, where the ore will be transferred to the canal. We would like to make a contract with you to carry this ore from Mead's Basin to Durham. If the price is reasonable, we will build a very large furnace at Durham.

The Midland Railroad, mentioned in Hewitt's letter, never came any closer to Ringwood than Pompton where its trackage went west rather than north to the Ringwood area. Their first train ran through Pompton westward in March of 1772. Just about this time the newly formed Montclair Railway Company was following through on plans to run a line through Pompton north up the valley toward Ringwood, but turning west to Greenwood Lake before reaching Hewitt's mines. By 1874 their trackage ran as far as Monks. It was this railroad, later known as the New York and Greenwood Lake Railway, that built a spur to Ringwood to make it possible for the first time for Hewitt to send his ore and iron products by rail, right from his ironworks property. Hyde's 1878 *Atlas of Passaic County* shows the completed spur line. Unfortunately the railroad came a little too late.

During 1872 and the first part of 1873, the market price of iron improved, but following the famous panic of 1873, hard times again returned to Ringwood that lasted well into 1879. Throughout these years, the Ringwood works were in effect inactive, although Hewitt stated in 1878 that the works had not been abandoned and that the furnaces could be put into blast whenever the market warranted it.[27] One of these furnaces was actually placed in blast late in 1879, but approximately six months later it was blown out, because of failure of the waterwheels. From then on most of the operations centered in Ringwood consisted of mining, with only the larger and more important mines supplying ore.[28]

Ringwood's production was steadily decreasing. By 1893, all mining had been discontinued and no Ringwood ore was sold at any time during this year. This suspension of operations caused a state of economic distress in the area. Workmen were not paid and a pessimistic air settled over the old iron center.

Five years later (1898) Hewitt wrote his former superintendent, P. R. George, that in the absence of mining operations Ringwood would become simply a place of residence without a future either for himself or others who still lived there. In 1900, however, the Peters mine was again being worked and continued to be active for a good portion of the time until 1931 when it was abandoned.

In 1942, the Defense Plant Corporation of the federal government bought the Ringwood mines believing there was ore there that could speed victory in World War II. The Allan Wood Steel Company was given a contract to dewater and recondition the

Skip car about to descend into Peters Mine. The ore-processing plant is shown in the background. *Photo by Raymond S. Darrenougué, 1952.*

Peters and Cannon mines. Old buildings which had been standing since Hewitt's time at the Peters mine were razed and new plants were erected. The total cost of rehabilitating these mines approached $4,000,000. Little—if any—ore was used from these mines during and after the war; the properties were sold on several occasions to various private companies that tried, but were not able, to make a commercially profitable venture out of the mines.

On February 28, 1958 the Pittsburgh Pacific Company of Crosby, Minnesota, purchased the works at government auction, taking title to the property July 16. Thus far, the new owner has not operated the mines and its only source of income has been limited to the rent on some 60-odd dwellings. Just what the future has in store for the old Colonial mines and works is not known, but a recent letter from E. T. Binger, president of the Pittsburgh Pacific Company, gives the company's present policy toward the properties:

> . . . Since the acquisition of this property the effect of foreign competition in the iron ore market has made it impracticable for this com-

> pany to activate these properties. Since the acquisition of the Ringwood Mines there has been no commercial operation of the property. Considerable test work has been done and until now the properties have been kept in a standby state. Pumping has continued in both the Cannon and Peters Mines but since we have no immediate plans for these properties, pumping will soon be terminated. Present plans for these properties include selling as much of the personal property and perhaps some of the real estate, as will not prejudice our future plans to operate this property for the production of iron ore. In other words, we plan to reduce carrying costs to a minimum so as to be able to hold this property for its probable iron ore values for some time in the future.[29]

There is, of course, one part of this historic estate which was not sold to the Pittsburgh Pacific Company. This is the picturesque site where the ironmasters had their homes and most of their forges and furnaces. On the east side of the roadway leading into the grounds at Ringwood is an embankment now covered with rhododendrons and evergreens, that once was the focal point of considerable iron-making activity. At the foot of this slope, 150 yards north of the bridge over the Ringwood River, stood one of the old furnaces, along with its waterwheel and casting house. The present roadway goes right over this spot.

Not long after he bought the property, Abram Hewitt dismantled the old furnace, placing the stone from the stack on the side of the embankment to form the present waterfall. The water that spills over this falls today comes from the millpond via an underground raceway, providing the power for the present little waterwheel located just south of the falls, next to the dairy building. Part of this building is believed to have formed the foundation and walls of Hasenclever's stamping mill. Portions of two old wheel pits can be found here—one on the north side of the building, now covered over by large stone slabs, the other under the south end of the structure.

The stable, located a few yards to the southeast of the dairy, is of more than passing interest. The old masonry walls in part of its basement may have been those of Hasenclever's forge of three fires and two hammers. Governor Franklin's committee, in their report made in 1768, stated that such a forge was within 50 yards of the furnace. This building lies just about this distance from the furnace site. This committee also mentioned another forge, located about 500 yards above the furnace, with four fires and two hammers. A discovery was made in the fall of 1961 of a forge site about the same distance upstream from the furnace on the west bank of the river.[30] Partial exploratory diggings have already uncovered the ruins of a hearth, as well as the foundation for a trip-hammer, plus other iron artifacts. Three other mounds, similar to the excavated one, are clearly visible—probably the locations of other hearths or forge fires.

The oldest building on the grounds is the small red house, the central part of which may have been the original residence of the Ogdens. It appears to have been built as early as 1740, the year they acquired the property. This home has an interesting history. During the Revolution it was used as a hospital, then for years as a Catholic Church, and still later as a gardener's home.

Hasenclever's manor house is no longer standing. It is thought to have been located in front of the present mansion. The current Victorian exterior with its many gables and wings is a relic of the days when the Hewitts occupied the house. The date of construc-

Ringwood Manor House. *Photo by Claire Tholl, 1960.*

tion of the present manor house cannot be determined absolutely, but Martin Ryerson probably built the original unit, the west end of the present house, within six years' time after acquiring the property in 1807.[31] In 1804 John Old, in advertising the Ringwood estate for sale in the *New York Herald*, stated: "There is on this estate an elegant Mansion House, 92 feet front, and about 30 feet deep, on which there has lately been expended a large sum of money, and is now in perfect repair." [32] Whether the house burned down or was demolished by Ryerson is not known. The east wing of the present mansion is actually a combination of three buildings, the appearance of which has been considerably changed due to rebuilding and remodeling over the years.[33] The house and grounds provide a suitable landmark for reminding visitors of the days when Ringwood held a widely-known position in the nation. They were deeded to New Jersey in 1936 by Erskine Hewitt, Abram's son.

Despite the nineteenth century appearance of this sprawling mansion and its formal garden, men connected with Ringwood's illustrious past—other than Ryerson, Hewitt and the more recent titans of America's industrial age—seem to lurk around the wooded slopes and banks. In the shadows surrounding the old casting bed and on the old road that winds down past the burial place of Robert Erskine and his faithful clerk, Robert Monteath, one seems almost certain to encounter Peter Hasenclever looking over the prospects for a fabulous iron empire, or Robert Erskine hurrying on his way to the old forge to oversee the casting of iron kettles for the Continental Army. Historic Ringwood is well worth a visit today with much of its past still on display.

CHARLOTTEBURG IRONWORKS

Less than a year after he had begun making iron on the Ringwood properties, Peter Hasenclever, furthering his dream of an iron empire, bought 6,583 acres on both sides of the Pequannock River about fifteen miles southwest of Ringwood. He called the new site "Charlottenburg" in honor of Charlotte, wife of George III.[1] His principal purchase of 5,619 acres where the furnace was erected was named the "Great Charlotteburg Furnace Tract." Records show that it was obtained on October 25, 1765 from Oliver Delancey, Henry Cuyler, Jr., and Walter Rutherford, agents of the East Jersey Proprietors. At the time of purchase, the Pequannock River was the boundary line between Morris and Bergen Counties; the larger portion of the tract lay on the south side in Morris County, the rest on the north side in Bergen County. Today the latter part is in Passaic County, which was carved out of portions of Bergen and Essex Counties in 1837.

Hasenclever devoted considerable time and expense to planning the Charlotteburg works. According to his account he started laying out, and possibly even constructing, buildings here before the dating of the deed. For example, Hasenclever stated that during the 18-month period following May 1, 1765, he built the following at Charlotteburg:

1 Furnace (not finished)
2 Forges, of eight fires, 80 by 45 feet long and wide, waiting only for the bellows before going into operation
1 Stamping-mill

3 Coal-houses	2 Saw-mills
2 Blacksmith shops	3 Stables
7 Frame-houses, with bricks	1 Carpenter's shop
37 Long-houses	2 Reservoirs
3 Store-houses	3 Ponds & 5 Bridges

The furnace, made of stone, stood on a bend of the Pequannock River and was the principal westernmost structure on the tract. Three miles downstream and east of it was the Middle Forge, equipped with two hammers and four fires for converting pig iron into bar iron. (The dam that supplied waterpower was "upwards of twenty feet high.") A mile downstream from the Middle Forge was the Lower Forge which had similar apparatus. Surrounding the forges at intervals were coal houses, dwellings for the workmen, storehouses, workshops, and stables.

Hasenclever improved transportation and waterpower facilities. He built a number of roads within the Charlotteburg area to provide easy access between processing points. Some of these were literally hacked out of the steep, rocky slopes that surrounded the works, while others in the low, swampy sections were built up with logs to make a firm footing for the horse-drawn carts and ore wagons. To assure the waterpower needed for operating the iron-works, Hasenclever raised dams to convert Macopin and Dunker Ponds into reservoirs.[2]

Hasenclever hoped that the works would be almost complete after his return from England. But when he came back in August of 1767, he found that two new coal houses had been burned, along with some new bellows planks. This accident held up any further work on the furnace for a period of time because, as Hasenclever pointed out, "such planks cannot be bought, they must be cut and dried on purpose." The furnace had not been completed when the committee appointed by Governor Franklin to appraise the American Iron Company properties arrived at Charlotteburg on July 2, 1768. But the committee noted that the furnace was almost finished, and called it one of the best blast furnaces it had ever seen in the Colonies. Great praise was also given to the buildings, roads, and reservoirs, because of the ingenuity in their design and construction. With such an abundance of ore and wood for making charcoal the committee estimated that the completed furnace could turn out 20 to 25 tons of pig iron per week at relatively small expense. Furthermore, the provisions made for an adequate water supply enabled the ironworks to operate throughout the year. In praising the Middle and Lower Forges at Charlotteburg, the committee quoted the overseer and workmen as saying that each forge was capable of making 250 tons of bar iron "single-handed" and close to 350 tons working "double-handed." This report on Charlotteburg shows great respect for Hasenclever's achievements.

The furnace was in operation by the latter part of the summer of 1768 and exceeded the committee's expectations. Among the Sir William Johnson papers in the state library at Albany, there is part of an interesting letter, written by Hasenclever in New York on September 30, 1768 to Johnson, mentioning the Charlotteburg works. Though badly damaged by fire, the legible portions which remain are significant.

> I have had this year 4 Furnaces in Blast, 3 of which . . . and 12 Forges, I have kept however a set of worcks under . . . 1 Furnace & 4 Forges, in order to Shew the difference between . . . & Capacity . . . these Worcks are Called Charlottenburg . . . is Carried on by my Germains & makes every week 28 to 30 Ton of Pigg Iron; When Two Furnaces under the Direction of the Agent whom the Company has send, make only 28 to 29 Tons together for which they Use 7560 Bushels of Shar Coal every week more & 50,400 lb ore . . . the People under my Direction at Charlottenburg do, my . . . will Soon Cry out Pater Peccavi and I hope to Convince them by evident proves [sic], that they have done me Unjustice . . . it is their duty to give me Satisfaction. . . .

It thus appears that in some way Hasenclever continued to control the works at Charlotteburg despite the fact that Humfray had been sent over from England the year before to take over all of the company's ironworks.

Hasenclever also proudly claimed that production at the Charlotteburg furnace was greater than that of the Hibernia furnace, which belonged to Lord Sterling. There they made about 17 tons only per week, as against 28 tons at his ironworks. Since both furnaces used the Hibernia mines as their source, Hasenclever stated that the difference in volume clearly showed the importance of a better-built furnace and an experienced, seasoned ironmaster. The latter was John Jacob Faesch, a skilled ironmaster Hasenclever had brought over from Europe in 1764 to supervise all of his ironworks in America. He had promised Faesch so much—large wages, a home, land, and other inducements—refusal was out of the question. Faesch not only had a thorough working knowledge of the furnace and forge operation, but was also able to pass this on to his workmen. Despite the fact that he was to oversee all the works, he apparently spent most of his time at Charlotteburg and it is to his credit that this furnace did so well. In fact, he seems to have had an even better and more practical understanding of iron manufacture than did Peter Hasenclever. However, Faesch left the American Iron Company's employ in 1772 to start ironworks of his own at Mt. Hope, New Jersey.

The Charlotteburg works had one significant advantage over Ringwood. By putting his waterwheel under cover and using stoves during the first winter freezes, Hasenclever was able to keep Charlotteburg in operation after Ringwood was forced to stop. When Robert Erskine took over the properties of the company after Faesch had left Charlotteburg, he praised the bellows and the quality of the ore supplied to Charlotteburg as being the best. Erskine's keen business sense told him that a promising potential existed at these works if only he could hire a larger number of workmen than was presently employed to haul the ore from the Hibernia mine to this furnace. As an inducement he offered to pay them by the ton according to the following seasonal schedule:

10s 6 d	If they started work before October 12th
10s	If they began after this date
9s	For winter hauling, when sleighs made the work easier
8s	For those who only worked in the winter [3]

An added stipulation was that each man had to haul not less than 3 tons per week until he had carried a total of 30 tons from mine to furnace.

Many writers in the past have stated that the Charlotteburg furnace was abandoned after 1772. Others have disputed this, placing the last working year as 1776 because of the destruction of the works by Loyalists. Dr. Joseph F. Tuttle, noted Morris County historian, wrote in 1869:

> It is a popular and widely believed tradition, that the English government, believing that the Americans were mainly dependent on the London Company's works for iron, made an arrangement with the Company to destroy them, in order to injure the Colonies in the difficulties which they were evidently approaching. It is very possible that some such proposition may have been made, but the only evidence I can find at any attempt to carry it out is in the destruction of the works at Charlotteburg, and the fact stated to me by some old men, that in the forests about those works, they have often seen coal-pits, which seem to have been burned down many years before, but the coal was not used, showing a violent suspension of business at some time. Those works were destroyed and the common belief is that it was done by direction of the Home Company. Still it must be admitted that the basis of the rumor is quite shadowy.[4]

Admittedly, there is no evidence to support these theories advanced by Tuttle and others. In fact, the only available evidence seems to contradict them. For example, the *New-York Gazette*, July 12, 1773, definitely established the fact that Charlotteburg was a going concern at that time: "Forgemen—A few good forgemen may hear of constant employment and sure pay by applying to the Foreman at Charlotteburg Iron Works, New Jersey. (N.B. Those who are Germans or can work in the German way shall be preferred.)"

The manuscript "Waste Book" kept by Robert Erskine, now in the New Jersey Historical Society, shows that bar iron was being received from Charlotteburg in July and August of 1774; in fact, more of it was produced by Charlotteburg in July of 1774 than either by Ringwood or Long Pond, the other Erskine properties. Additional data indicates that Erskine equipped and drilled groups of men which he stationed at the various works under his managership for protection against possible enemy attacks in 1775. In a letter to Lord Sterling written in May of 1775, Joseph Hoff, Manager of the Hibernia furnace, specifically mentions Charlotteburg as one of the places where a unit of this type was located.

Other documentation that proves Charlotteburg was active during 1776 and 1777 are letters from Erskine to George Clinton, Brigadier General in the Continental Army and Governor of New York. In one of these Erskine complained that the lack of charcoal was hindering his efforts to make iron for the *chevaux-de-frise* ordered by the army. (This obstruction was to be placed across the Hudson to deter British warships from advancing up the river.) Erskine said further that unless his woodchoppers returned, he could not again "blow his furnaces" because he lacked enough charcoal. He estimated that his supply would last only to the middle of May and his stock of "pigs" about five months. In another letter to Clinton written March 3, 1777 Erskine said: "All hands in general are gone off from the woods; I have but two remaining at Ringwood, seven at Long Pond and half a Dozen at Charlotteburg . . ."

The manpower shortage caused Erskine to appeal to General Washington himself. On February 27, 1777, he asked the general to exempt most of the men at the ironworks from military duty, pointing out that "his neighbors Coll. Ogden & Mr. Faesch had obtained an exemption for the hands employed at their works." But Washington refused the request, saying, according to Erskine, that "he had met with some impositions in that way already, and indeed, justly observed that all Ironworks carried on for private emolument might demand the same favour. . . ."

Had there been any serious threat of attack, it is highly unlikely Washington would have turned down a request of this sort. No doubt the general was well aware that articles of paramount importance to the success of the Continental cause were manufactured at these ironworks.

Erskine's request to Washington strongly implies that all the Erskine-controlled works, including Charlotteburg, were in operation at that time. Evidence that the Charlotteburg works were in existence well after the outbreak of the Revolution, is found in a letter to Clinton written March 14, 1777: "The Bearer, Mr. Ambrose Gordon, whom I can amply recommend for his honour and attachment to the cause of America is very solicitious to serve in the new Levies. Mr. Patrick Hayes likewise, whom you know, and Mr. William Harrison, to whom I intrust the Care of Charlotteburg, have both an inclination for the service; but as I cannot part with them at present, I do not now recommend them; though I could do it with equal Confidence and justice in every respect. . . ."

Ebenezer Erskine's diary gives further proof. In September, 1778, he wrote that Robert Erskine still kept a clerk at Charlotteburg, although the ironworks there, and those at Ringwood and Long Pond "had mostly stopped as he now had only about 40 hands employed [whereas] he used in former times between 3 & 400 in constant employ." Since there is no further mention of Charlotteburg in the correspondence of the Revolutionary years, it is most probable that the manpower shortage caused the works to be shut down. The only other reference to Charlotteburg during the eighteenth century occurs in the vivid but dubiously accurate travel accounts of Hector St. John de Crèvecoeur:

> We arrived at Charlottenbourg across a very mountainous country. The constructions there had been built before the Revolution by an English Company which the war had ruined. The furnace had just blazed forth. The proprietor was absent. We saw an immense nail manufactory extremely simplified by reason of a great number of small hammers put in motion by an exterior shaft. Workmen were forging bolts, also several other articles of iron for use on boats. We

saw also a flattening process for sheet-iron and iron blades, used in the making of spades and shovels. They told us that last year they had melted 46,000 hundred-weight of pig iron. There, as at Sterling and Ringwood, the water reserve was immense.

During the latter part of the eighteenth century and the nineteenth century Charlotteburg changed hands several times. John Travis, a Philadelphia merchant, acquired the ironworks by deed on February 6, 1796 from William M. Bell, high sheriff of Bergen County. Nine years later, on May 11, 1805, Travis sold the Charlotteburg property—almost six thousand acres—to Judge Elisha Boudinot of Newark. On September 1, 1807 Boudinot sold half of the tract to Martin Ryerson for $3,406. It is probable that the once-busy hammers and bellows of the forges and furnace continued their thirty-year decline as there is no record of Ryerson's having used them during his lifetime.

In 1840 activity began again at Charlotteburg when George H. Renton, who had been associated with the iron industry in Newark, built a bloomery and rolling mill on the site of the old furnace. Renton employed C. F. D'Camp as his manager for the rehabilitated works. The bloomery had six fires and two hammers; the mill had two heating furnaces and three trains of rolls. (The iron was run between trains of revolving rolls and came out in flat sheets.) Both works were driven by waterpower. D'Camp utilized the escape heat of the forge fires to heat the blooms for rolling, thus efficiently reducing the amount of anthracite needed for the operation. In the October, 1853, issue of *The Mining Magazine*, a brief but comprehensive report summed up the lucrative possibilities of Renton's business venture at Charlotteburg: "It will not be far out of the way to say that some fifteen men are usually employed in this mill, and that 500 tons of finished iron are rolled annually, using 650 tons of blooms, worth $27,000. The finished iron is worth $42,000. The labor to produce this may not be far from $4,000 and the extra fuel for the furnaces about $1,000, leaving a handsome profit to the capitalist."

The old Charlotteburg mine, believed to have been first opened about 1771, was worked intermittently during the eighteenth and nineteenth centuries. There are records of ore taken from it during

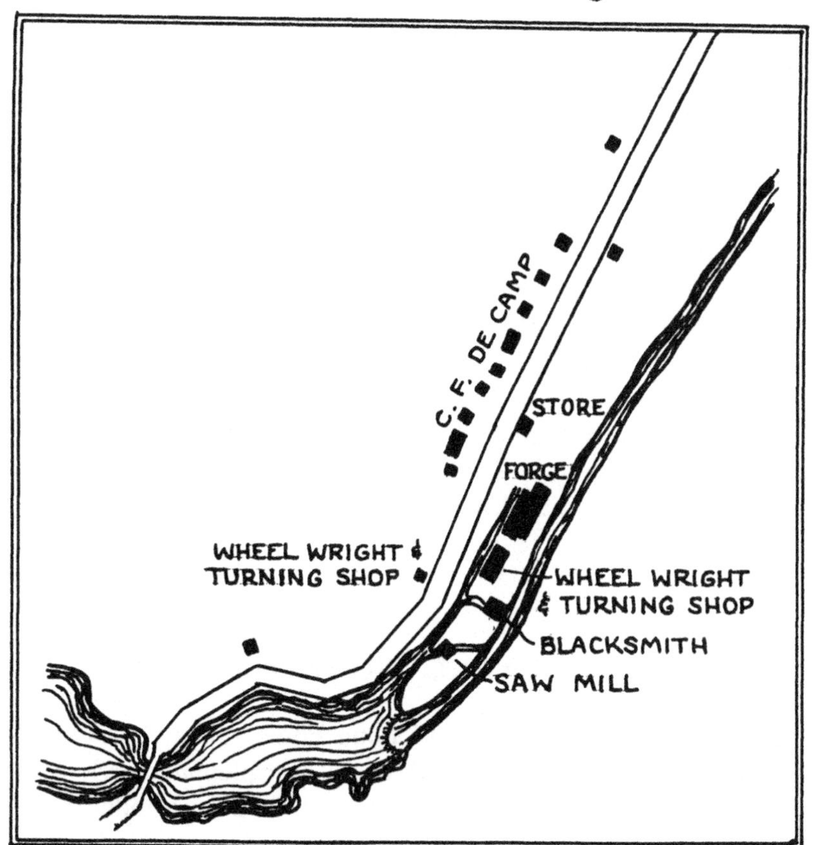

Charlotteburg. From Corey's wall map, Philadelphia, 1861.

the 1840's and 1850's but it apparently closed shortly after. In 1872, when it was reopened by the Bethlehem Iron Company, its depth was reported as 60 feet. During the panic of 1873, the mine was shut down again. A few years later it was reopened and two new veins were worked. It was idle in 1879 and worked again from 1880 to 1884.

The furnace area with the scattered remnants of the once-flourishing Charlotteburg works is now beneath the waters of a new reservoir, much larger than any Hasenclever ever constructed. As recently as 1960, a visitor to the furnace site could still locate the old raceway, waterwheel pits, a very old dam, a more recent one, and the foundations of former buildings that had crumbled to the ground.[5] A fractured tie rod, picked up here and carefully examined to determine its age, was found to date back to the Colonial era and doubtless was one of those used in the furnace built by Hasenclever in the 1760's.[6]

Of great interest is the recent discovery of the site of the old Middle Forge by Frank Malone of Oradell and James Norman of Newfoundland, New Jersey. They and other members of the North Jersey Highlands Historical Society have already uncovered an old sluiceway, waterwheel pits with beams preserved below the water level, traces of a charcoal house, three hearths with marks of forge fires, two hammer bases, and even a piece of "pig" dated 1770. This site, which is not included in the new reservoir, promises even more interesting "finds," as excavation work goes along in hope of shedding further light on Charlotteburg's historical past.[7]

LONG POND IRONWORKS

The crumbling ruins of the old Long Pond ironworks lie on the Wanaque River (the Long Pond River in Colonial days), about three-and-a-half miles west of Ringwood, and two miles downstream from Greenwood Lake.[1] When Peter Hasenclever began to build Long Pond in 1766, his first thought was to assure the works of a constant supply of water. To that end, he constructed a dam, 200 feet long and over five feet in height, across the southern end of the lake. The same year, he began building "1 forge with 4 fires and 1 furnace, 2 coal-houses, 2 blacksmith shops, 4 frame-houses, 6 log-houses, 1 store house, 6 colliers houses, 1 saw-mill, 1 horse-stable, 1 reservoir, 2 ponds and 2 bridges," all to be linked by roads. Steep mountain slopes, swamps, and gullies made the road-building costly.

The furnace was not completed until early 1768 but by July it was in blast with a capacity of from 20 to 25 tons of pig iron per week. The furnace at Long Pond was considered superior to those at Ringwood and Charlotteburg because of its elevation, which helped keep the iron "free from damps, etc." Although other mines were closer to Long Pond, the ore used in the furnace was carted

Long Pond Ironworks. *Photo by Vernon Royle, 1909.*

from the Good Hope mine at Ringwood because of its excellent quality.

At Long Pond, Ringwood, and Charlotteburg, Hasenclever introduced such iron-making innovations as furnace walls of slate, overshot waterwheels, and hammer wheel shafts equipped with strong cast-iron rings whose arms served as cogs to lift the hammer handles. Because of their close association, it is impossible to discover at which one of these three early ironworks such improvements were first introduced.

As noted previously, Hasenclever was succeeded as manager of all the American Company properties at Long Pond, Charlotteburg, and Ringwood on October 1, 1767, by Jeston Humfray, who in turn was succeeded by John Jacob Faesch. Robert Erskine and Faesch managed the properties until the middle of the summer of 1772, when Erskine became sole manager of all, keeping the job until his death in 1780.

Just how long after the outbreak of the Revolution Long Pond was in use is not known, but the entry in Ebenezer Erskine's diary in 1778 noting that the ironworks had almost completely stopped for lack of manpower indicates it was not in operation throughout the war.

For the next twenty years the Long Pond works remained idle, although the property changed hands several times. Records show that John Travis bought the works on February 6, 1796, and sold them to John Old on January 1, 1798. Old put Long Pond up for sale in 1803, advertising "that Long Pond ore goes 27 miles to Demarest Landing, Hackensack, N.J." This had also been the shipping point for Long Pond iron during the Colonial period. Apparently no better route for shipping iron from the Ramapos had been discovered since pre-Revolutionary times. The cartage from Long Pond to Demarest Landing (now River Edge) involved a long and circuitous journey, particularly hazardous from December to April.

Martin Ryerson acquired Long Pond when he bought the Ringwood properties in 1807. Although information is available about operations at Ringwood during this time, little is said of Long Pond. An inventory taken in 1840 after Ryerson's death the previous year lists:

7	tons of bar iron at Long Pond Forge	@$75.	per ton	$525.00
12	pairs of tongs, scales, weights and implements at Long Pond Forge			35.00
2	crow bars and one cant hook, 1 axe			3.00
164	boxes of coal at Long Pond Forge 18,040 bushels	@.05	per bu.	902.00
3	mules, one of two years old			125.00
20	tons of hay in two barracks at Long Pond swamp	@$9.	per ton	180.00
	Lumber at Long Pond Saw Mill			25.00

This relatively brief mention of the Long Pond forge, without any reference to the furnace or supplies of ore on hand, not only implies that the furnace was idle, but that it probably had not been operating for some time. It contrasts sharply with a contemporary inventory of the Pompton ironworks which clearly indicates that production was being maintained there at the time Ryerson, who owned both works, died. Martin Ryerson willed the "Long Pond Great Furnace Tract" to his son Jacob, who also inherited the "Ringwood Great Furnace Tract."

A new period of activity for Long Pond began when the Ryerson heirs sold it, with Ringwood, to Peter Cooper and Abram S. Hewitt of the Trenton Iron Company, on September 1, 1853, for $100,000. The partners probably began to work the forge soon after acquiring it, although the exact date is not known. Business was very bad, as it was in general, but eight years later, the guns of Fort Sumter signaled a new era for the iron industry. Hewitt and Cooper intended to develop Long Pond into a major iron production center:

> [Hewitt to Mr. Abbott at the Trenton Iron Company, October 30, 1861]
>
> A new Charcoal Furnace is to be erected at Ringwood [Long Pond]. Do you have a man to do the work . . . a good founder to run the Furnace?
>
> [Hewitt to P. R. George, superintendent at Ringwood, November 27, 1861]
>
> Peter Cooper, Edward Cooper & Mr. Edmund R. Miller will come up Friday to settle the Furnace business.[2]
>
> [Hewitt to Charles Knapp, Washington, D.C., December 4, 1861]
>
> As to Ringwood, the men are digging out the foundation at Long Pond, but if you decide to take the contracts anyhow, I think we shall do better to build the Furnace at Ringwood, as originally proposed. We shall not get iron so quick, but the whole affair will be more satisfactory when done. I suppose that we shall be able in the latter case to give you iron by May 1st next. If we go on at Long Pond, we can do it by 1st of March, but in the one case we have the permanent work, and in the other a temporary job. I would like to know what you think best.
>
> . . . I believe we shall be able to give you what [iron] you require and on the whole prefer to spend $10,000 at Ringwood and have the thing done right, rather than have a poor thing at Long Pond.

Hewitt was not the only one trying to get both Long Pond and Ringwood on a footing to produce iron as quickly as possible. Edward Cooper, son of Hewitt's partner and also a close friend of

Hewitt's, wrote on December 16, 1861 to "W. Keyler" of Haverstraw, New York, making inquiry about "bottom stones for the Charcoal Furnace we are building at Ringwood [Long Pond] Passaic County, N.J." At least a part of this action was stimulated by the effort of the federal government to obtain artillery mounts and carriages as well as other badly-needed ordnance. In a letter written December 5, 1861 to Charles Knapp in Washington, Hewitt admitted that enterprising agents, aware of the Government's lack of arms-producing facilities, were bidding for the now-promising Ringwood-Long Pond properties: . . . "I do not believe that we could have iron from Long Pond by February. . . . I may as well add that a speculator has already applied for the price of Ringwood, with a view to offering it to the Government for a national foundry. We have, of course, given no encouragement to any such ideas."

From a letter written the day after this to P. R. George by Edward Cooper it is quite evident, that despite Hewitt's desire to have the new furnace built at Ringwood, he was forced to yield, presumably to the Coopers, for the letter reads: "We have concluded to go along with the Long Pond Furnace as proposed before I stopped you on Wednesday. Will be up next week and bring plans for the stone work." By January 21, 1862, about six weeks later, Cooper wrote to Watson & Company, at Perth Amboy, New Jersey, inquiring about the availability of bricks to form "an arch flat on the upper and under surface." He outlined their needs: "Enclosed we send you a tracing giving a section of the lining; boshes & hearth of a charcoal furnace we are now building. We want to keep the bottom of tymph 25 inches above the bottom of the hearth and the bottom of the tuyere arches 18 inches above the bottom of the top of the tuyere openings." Cooper mentioned that they would need 12- or 14-inch bricks for the furnace lining and sought the price per thousand bricks "on board vessel at Amboy."

The war with its constant need for arms and equipment made of iron undoubtedly was responsible for speeding the completion of the new Long Pond furnace. In a letter, September 12, 1862, Abram Hewitt said: "We have now got a charcoal furnace at work at Ringwood, making very superior iron." It was necessary to have the best equipment to get profitable contracts from the government for making supplies for the Union Army. Hewitt was well aware of the importance of this new furnace to the success of the enterprise at Ringwood and Long Pond. On September 17, 1862, he wrote to the South Boston Iron Company about the good results being obtained from the new furnace: "We have purchased a new steam hammer . . . the pig iron being turned out at the Charcoal Furnace recently put in blast on our Ringwood property is very strong and supposedly a superior gun metal."

In a letter to Ringwood, January 5, 1863, Hewitt impressed on his superintendent the necessity for turning out only the best possible grade of iron: "Nothing but No. 1 Ringwood Foundry Pig Iron will answer for gun-barrels and the furnace must therefore be run so as to make it. If necessary, the blast must be raised when the furnace is inclined to fall off."

Throughout the remainder of 1863, frequent references detail the business of running a prime producing furnace. In May, Hewitt, ever watchful, complained to his superintendent about the rising expenses:

> I am puzzled by the Ringwood expenses. During the six months ending Dec. 31, 1862, we paid out over $39,000, and as the furnace was completed in the first part of the six months, I suppose that $6,000 per month is a fair average of the regular expenses. At any rate, we have paid about $6,000 per month since January 1st and if we go on at this rate, we shall pay out $72,000 per annum.
>
> You make 1,500 tons of pig and 300 tons of blooms in a year, which is an average of $40 per ton. Now the pig ought not to exceed $25 and the blowers $35. What becomes of the rest of the money? I am afraid there is a big leak somewhere, and I think it must be in the wood and coal.

During the Civil War years, naturally, there was a scarcity of employees which hindered operations. In the same month he had cautioned about the waste of money at the properties, Hewitt sent an encouraging note to the company headquarters at Trenton about a new source of badly-needed manpower: "There are plenty of emigrants to be got now, good strong men. We are getting them for Ringwood and Long Pond at $15 per month and board."

Another problem—shipping the surplus ore (too much to use at Long Pond) from the Ringwood mines—was solved in 1862 and 1863 by having it carted 20 to 25 miles to Piermont, New York, on the Hudson River. There, the ore was loaded on coast-wise schooners and shipped down-river into the Atlantic, around Cape May into Delaware Bay and up the Delaware River to Trenton, where it was used in the Trenton Iron Company's furnaces.

Immediately following the Civil War, Hewitt apparently was willing to sell the Ringwood properties, including Long Pond. In one letter, he told the Bessemer Steel Works Company of Troy, New York, which must have expressed an interest in buying, "Ringwood pig iron is made from magnetic iron ore of great purity, yielding about 65% in the furnace." He said that the "only impurity is silica, no sulphur or phosphorus. There are over twenty beds of iron on the property and two blast furnaces, capable of producing about 100 tons per week.[3] The blast is usually between 300 and 400." But the properties were not sold. Market conditions took an upturn in 1868 and Hewitt bought out one of his partners as he told P. R. George, April 15, 1868: "I have purchased Mr. Miller's interest in the joint property and desire to run both furnaces. Put everything in running order immediately and have ore taken out . . . the hematite to Piermont. The Sterling Railroad cars will take it." [4]

Three months later, still eager to take advantage of the current market, he was looking for a founder who would be willing to fit into the position which he described as "conducted in rather an old-fashioned way, and the wages are not high. The privileges, however, are considerable."

Another sign of the revival in activities at this time is an order written by Thomas Hewitt on July 13, 1868. Thomas, a relative of Abram, was then acting as manager at Long Pond. The order, as listed in the old Long Pond furnace ledger, was for bar, square, round, hoop, and hammered iron; cast and blistered steel, files, rasps, chains, molder's shovels, and a large supply of letterheads and envelopes, as well as food and clothing for the workers and their families. About five weeks later, another letter noted "three teams . . . loaded with No. 1 iron going to Trenton . . ." Further evidence from correspondence of the time pointed up a need for "Iron shovels" and coal "as we are entirely out and suffering for it." [5]

Writing to the superintendent on November 30, Thomas Hewitt boasted that during the week of November 22 to 28, a little over

Office and store at Long Pond. *Photo by William F. Augustine, 1957.*

61 tons of iron had been manufactured at Long Pond. The iron, he said, was composed of one-fifth hematite—and the balance was three-quarters "Hard Mine ore" and one-quarter "Cooper Mine ore." Subsequent reports for the following weeks failed to reach the same production levels but averaged close to 54 tons per week. Perhaps the best indication of the manner in which Long Pond was working,

now that it was in full production, was the summary of January operations in which a total of 207 tons of iron were turned out at Long Pond, of which 159 tons were No. 1 grade.[6]

In spite of the apparent success of Long Pond, Abram Hewitt seems to have been plagued by excessive operating costs and very low profits which made the venture a risky business proposition. He complained about the financial problem in a letter to Mr. George on February 27, 1869: "There is a loss of $26,000 on the operations of 1868. The pig iron has cost about $50 per ton. I am determined to ascertain what has become of this enormous sum. At present we are all in the dark." Some of the figuring that went into iron mining and manufacturing in the nineteenth century appears in a letter from Hewitt to H. C. Cotton of Reynolds & Company of New York City, March 10, 1869. This rundown seems to have been a response to questions posed by Cotton as to the operational cost and methods of running a mine, forge, and furnace, as well as the advantages and disadvantages of steam versus waterpower:

> The cost of a blast furnace will necessarily vary with different locations and the cost of materials. Our Furnaces at Ringwood [Long Pond] with a capacity of about ten tons per day have cost about $50,000 each. They are driven by water power. If steam had been used, they would have cost about $60,000 each, or $10,000 more for each furnace.
>
> The question of running by water or steam is one purely of calculation with reference to the particular locality, and the projector can readily decide whether it is better, in view of the saving in cartage, to use steam, or in view of the saving in capital, it will be better to use water. We estimate however that the absolute economy of a water power, over a steam engine, is fifty cents per ton of iron made.
>
> At Ringwood we use fully two hundred bushels of coal to the ton of iron, at Lake Superior we use one hundred and fifty. I do not believe that charcoal can be made and delivered to the furnace at an average distance, two loads per day, for less than seven cents per bushel. At Ringwood the cost of iron is thirty-five dollars per ton, at Lake Superior twenty-five dollars. As to capital, we use about $100,000 for each furnace.

No records have been found which give the amount of iron produced and shipped from Long Pond after March 18, 1869. A brief reference is found in a letter written by Thomas Hewitt on October 14 which merely states that "the Iron we forward you today commencing with Chas. Bishop's second load has no hematite ore in and was made with a moderate heat stamped No. 1." The last records showing shipments of iron from Long Pond are those of October 19, 1869. But the Long Pond furnace continued to turn out iron, despite the cessation of shipments from the area.

Among the interesting data compiled in connection with the industry at this time is a typical month's payroll which shows how large a portion of the salary earned by the average employee was taken out in trade rather than paid in cash. In August 1869, according to an old ledger, furnace workers' salaries were $1,591.84; store "help" $100.10; teamsters $383.65; and coal burners $2,119.18.[7] Almost three quarters ($3,566.05) was paid in the form of goods, supplies, and merchandise, while only $628.72 was in cash.

As far back as May, 1869, Hewitt had evidently been approached in regard to selling the mines in and about Ringwood but he was reluctant. In answer to a query from Alexander Ellicot of Dover, New Jersey, who probably was a real estate agent, Hewitt replied on May 18 of that year:

> The matter of selling the Ringwood Mines requires a great deal of consideration, but I might be persuaded. . . . We have mine rights on twenty thousand acres, covering all the ranges of ore from the most easterly line to the range of the Ogden Mine. The length of the property is nine miles and it would take weeks to see all the places where ore has been found. . . . I will pay you a two and one-half commission if I decide to sell. The extent of this property is so great and the quantities of ore so immense, that I find it very hard to come to any decision, and I certainly would not entertain the idea, even at a million dollars except for my wish to get out of all business while I am young enough to recover my health, etc.

From Hewitt's words, it appears that he was not over-eager to find a buyer, but because the properties were not producing as high a profit as he estimated they should, he was "entertaining" the thought of selling. Whether Ellicot ever made an offer, either on behalf of clients or for himself, is not known. But Ringwood and

Long Pond Ironworks, circa 1900.

its related properties, including Long Pond, were not sold at this time.

In what must have been an effort to solve the problem of waste which was costing him huge sums of money at Long Pond, Hewitt installed a new type of kiln there in 1868. The first mention that can be found of this device, which seems to have aroused a great deal of interest in the industry, is in a letter, November 12, 1869, to Isaac Butts of Rochester, New York.

Hewitt pointed out: "We have installed a Westman Kiln, that will remove all sulphur. It is operated by the Waste Gas from the Blast Furnace. It cost about $10,000. You can see it in operation at any time in Ringwood [Long Pond]." About a month later, in a letter to James Woods of the Cumberland Iron Works in Tennessee, Hewitt gave the only indication of the date when this new kiln was introduced at Long Pond. "The Westman Kiln has been in operation for nearly 1½ years with success, roasting ore with the waste gases from the Furnace exclusively, and with far greater uniformity, than in kilns." [8]

A year later, in spite of the success of the new kiln, Hewitt found his business seriously affected by the drop in market prices as shown in his letter to the Swedish engineer, Sjoberg, who had built the

Westman kiln at Long Pond. "If the furnace is to remain in blast, the most rigid economy must be practised and every possible savings made. It will be too late after the thing is stopped, and I must close it up, unless we can prevent the actual loss which is taking place. And all hands will be out of employment, and there will be distress, which I want to prevent."

Long Pond furnace must have been taken out of blast about this time, or perhaps shortly after as Hewitt declared on May 15, 1871 that the furnace would "go in" blast the following week and that the famous Westman kiln, which was still an object of curiosity, could then be seen in action. In the same year a significant decision was made regarding a changeover in fuel. Hewitt had heard of increased efficiency by substituting anthracite for charcoal and he decided to rebuild the Long Pond furnace so that it could use the modern fuel.

By 1872 the work of reconversion was under way. This involved considerable renovation and enlarging of one furnace. The dimensions were increased so that it measured 48 feet in height, 13 feet in width across the bosh, and 33 feet across the square base. The other furnace (65 feet high, the bosh 16 feet wide, with a 35-foot square base) was cut down to a height of about 20 feet, but was never completely rebuilt.[9] It still stands today with its sawed-off appearance while its counterpart, although rebuilt, has entirely crumbled and is merely a pile of rubble. Hewitt also planned to build a huge waterwheel measuring 50 feet in diameter at the Long Pond works. The foundation for this immense wheel is still visible about 200 feet north of the ruins of the two waterwheels adjacent to the furnace site in Hewitt, New Jersey. Its large U-shaped walls formed by massive blocks are 14 feet high and 12 feet thick, with the entire foundation measuring 66 feet long and 23 feet wide.

During March of 1873, Hewitt noted in a letter that Sjoberg was at Ringwood, although the furnace was not in blast.[10] Presumably the engineer had been summoned to Long Pond to assist in putting the furnace and its kiln in active service again, but the furnace was not put in blast. Four months later, Hewitt told Joseph C. Kent of nearby Phillipsburgh that he needed "a good founder at Ringwood [Long Pond]" for the smaller of the two furnaces there.

Long Pond Ironworks. *Photo by C. Durand Chapman, circa 1894.*

> The Furnace is small, 12-foot boshes and the pressure of the blast does not exceed 3 pounds. She was badly blown in, but never-the-less showed she would work perfectly well. We use the old water-wheels, but have a new and powerful wheel underway which we hope to have completed by 1st January, when we will have all the blast and all the pressure the furnace will take. Now if your man will be satisfied with so small a furnace until No. 2 which will be a big one is ready, say next summer, then I would like to have him. The wages of course are too high for so small a furnace, but perhaps he will take less until the large one is ready. The [small] furnace is all done ready to start, but the Railroad is not yet done. By October 15th it should be ready.

Once again, Hewitt's hopes and plans for the Long Pond furnaces were frustrated. Despite the confidence with which he had outlined the prospects of the furnaces to Kent, in a subsequent letter to John Downs of Stanhope, New Jersey, during September, 1873, he admitted flatly that his ambitions had suffered a setback: "The failure to get the right to furnish the Railroad this year and the financial panic will both prevent us from putting the Ringwood [Long Pond] Furnace in blast at present." And not until the latter part of 1879 was iron production resumed when one furnace was started and worked for a short time. Hewitt was quite unwill-

he two waterwheels at Hewitt, New Jersey. *Photo by Charles F. Marschalek, 19*

Pipe conveying water to the wheel. *Photo by Vernon Royle, 1909.*

ing to classify the works as being "abandoned." He made repeated efforts to get either one or both furnaces back into action. Excerpts from further correspondence during 1880 show his increasingly gloomy view of conditions:

> May 13, 1880—Condition of Furnace is certainly deplorable.
> May 28, 1880—The Ringwood [Long Pond] Furnace has finally gone out of blast in consequence of the failure of the waterwheels and our want of success in getting one of them rebuilt in time. We got within 24 hours of saving it.

According to the *Directory to the Iron and Steel Works of the United States 1888* (corrected to November 1887), the Long Pond anthracite furnace was inactive and "had not been in blast for several years." Verbal corroboration of the approximate time at which this occurred was obtained over 20 years ago in the 1940's during a conversation with John Townsend, a former employee at the ironworks. The last reference to be found in Abram Hewitt's correspondence dealing with Long Pond is in a letter of October 14, 1887. Obviously the enterprising Hewitt was still wishing that he could put the Long Pond furnace into blast once again:

> We have a Furnace at Hewitt's Station which could be put in blast, if we could get back the cost for the iron which it would make. I am satisfied that it will not be possible in the present condition of the market to make iron sufficiently low in cost to warrant the putting in of the Furnace even if the price of Coke would not exceed $4 a net ton at the Furnace. Even at that price it would require a little investigation as to whether we could put the Furnace in blast and it would require two months time to get it ready.

The important problem of profit and loss was the deciding factor in Hewitt's reluctant decision to abandon, once and for all, the Long Pond site. Now these two furnaces would join other deteriorating structures at the works such as the forge constructed by Hasenclever in 1766. There was one difference, however. Today all trace of the early forge has vanished but the remnants of the two furnaces still remain. The well-preserved ruins of a third furnace, which appears to be very old and may have been the one first put into blast by Hasenclever (1768), have been discovered here

One of the two waterwheels at Hewitt. *Photo by Charles F. Marschalek, 1952.*

recently.[11] This furnace was unearthed in a mound lying 120 feet west of Hewitt's sawed-off furnace. During the days of May 6–10, 1963 Roland Robbins, a well-known archaeologist, did some partial excavating around the newly discovered furnace ruins disclosing a 26 feet square base and the type of lining—slate—that Hasenclever himself introduced in this country.

Two wooden waterwheels also survive but they were badly damaged by a fire set by vandals during June of 1957. Before this they were in almost as good condition as they were the day Hewitt closed the works. Blackened and skeleton-like, they are the only furnace wheels still visible in either southern New York or northern New Jersey. Each wheel is 25 feet in diameter with spokes a foot thick; the shaft 3 feet in circumference; the paddles or buckets had been 7 feet long, 14 inches wide, and 1½ inches thick. The width of each wheel is a full 8 feet, and the overshot type.

The ruins of the Long Pond ironworks are still interesting to visit and it is possible to envision the works with waterwheels, furnaces, and the raceway running north to south on the hillside, dividing behind the wheels into two separate passages, one for each wheel. By following a trench north and passing the huge stone foundation, which was never completed, a hundred yards farther one comes to the top of a rocky ravine with a waterfall. Just above these cascades there are iron spikes in the rocks, evidently used in constructing a former dam. This was probably one of several dams used over the years to raise the water level to the height needed to fill the raceway when Long Pond was active. Here the visitor can look down into one of the loveliest glens in the Ramapo Mountains. The combination of waterfall, overhanging hemlocks, and unique rock formations creates a landscape of charm and beauty.

BLOOMINGDALE FURNACE

The Bloomingdale furnace was erected about 1761 by John and Uzal Ogden on the south bank of the Pequannock, about 700 feet downstream from where Stony Brook empties into the river.[1]

John and Uzal, members of the Ogden family who had owned Ringwood, acquired a tract of land at the Bloomingdale site. It consisted of a little more than 137 acres, conveyed to them by Philip Schuyler and his heirs on August 1, 1759. In October, 1765, they added to their holdings by purchasing 34 acres from Guilliam Bertholf.[2]

Their acreage at Bloomingdale lay on both sides of the river. Among the natural advantages of this property were good waterpower, an abundant supply of timber for charcoal, and a sizeable, well-cultivated farm. The location was especially favorable because the Passaic River, only a few miles away, offered a convenient means of transporting the iron to market at relatively low cost.

Despite these advantages, the Bloomingdale furnace was not profitable. In 1770, Uzal Ogden made public admission of his insolvency and two years later the works was advertised for sale in the *New-York Gazette and the Weekly Mercury* of June 22, 1772. The sale notice stated that the Blast Furnace and all related buildings and accessories, late the property of John and Uzal Ogden, would be sold at the Coffee House in New York City. Interested parties were asked to get in touch with either Hamilton Young, a New York merchant, or Isaac Ogden in Newark.

The Bloomingdale ironworks was described in this advertisement in such promising words that it seems almost impossible it had closed after a short and unprofitable period of operation. The furnace was said to "afford more certain prospect of profit to the purchaser than most furnaces in America, especially where the quality of pigg metal that has been and may be again be made at it, is considered." The iron made here was described as particularly adaptable for use in air furnaces and, except for pig iron from one other locality, was the only metal "in this part of America that will answer for that purpose—as such the piggs are in great demand now, and will continue with a ready sale, and command the highest price."

The advertisement also called attention to the fact that the furnace and all buildings were in good condition and ready for business. The woodlands were thought to be capable of supplying the works for years to come; and it was claimed that the farm would furnish sufficient "bread corn" for a good-size family as well as forage and grain. When the sale was held, or whether anyone bought, is still a mystery. The furnace most likely continued idle even through the Revolution. It is shown on Erskine's 1778 map (90a) in the New York Historical Society.

On January 29, 1795, William Ellsworth acquired the "furnace and Iron Works at Bloomingdale." It is doubtful if the furnace was ever worked by him. Benjamin Roome, a surveyor, mentioned that it still was in fair condition in 1812 and that as far back as 1800—at least—it had not been in blast. By the early 1820's it was in ruins [3]

Ruins of Ryerson's Forge, Bloomingdale, New Jersey. *Photo by Vernon Royle, circa 1893.*

and on April 12, 1833, Ellsworth's daughter Henrika Leary, a widow, sold the Bloomingdale property to Martin I. Ryerson.

On March 11, 1872, the first train of the new Midland Railway, on its way to Newfoundland, New Jersey, passed within a few yards of the old furnace site. The ruins, recently uncovered, lie south of the tracks in the midst of a brush-covered mound on the grounds of the old Pequannock Coal & Lumber Company.[4]

BLOOMINGDALE FORGE

This forge was located on the north bank of the Pequannock River at Bloomingdale, New Jersey, a short distance downstream from the old furnace. The forge, built about 1800, was rebuilt in 1839, and again in 1841.

One of the earliest owners, if not the original one, was Martin Ryerson, proprietor of the Ringwood, Long Pond, and Pompton ironworks. He had willed the "Bloomingdale forge lot and the works thereon" to the children of his son John, deceased at the time the will was made in 1833.

After his grandfather's death in 1839, Martin John Ryerson (1814–1889) took over and ran the forge for many years. He eventually bought up the rights of the other grandchildren, becoming sole owner, as well as proprietor.

The forge continued to use Ringwood ore, principally, because provision had been made in the same will that he and his brother Peter M. Ryerson should have:

> the full right and privilege of taking any quantity of ore from the Ringwood Great Furnace tract for any works now erected or which may hereafter be erected by them or either of them with the full right of free ingress and egress at all times with their carts, wagons and teams, servants, agents and laborers to dig for and take away the said ore.

Waterwheel and bellows house, Bloomingdale Forge. *Photo by Vernon Royle, circa 1893.*

Trip hammer at Bloomingdale Forge. *Photo by Vernon Royle, circa 1893.*

In 1855 Martin J. Ryerson made 255 tons of bars and faggot iron for shafts and boiler plates from Ringwood ore. Besides running the forge, he also lived there and operated a sawmill, a gristmill, and a store. In the later years the forge had four fires and two hammers driven by waterpower.

Old photographs from the late 1800's show the waterwheels which supplied power to the shears for cutting scrap and bar iron, as well as the iron shaft with tripping mechanism to operate the stamp.[1] The trip-hammer and its anvil, and the chain supporting the tongs holding the hot iron, while being forged, are in plain sight. The cutting shears appear more modern, compared with the age of the wheels.

Bloomingdale was still active in 1879, but by 1882 Martin J. Ryerson had shut down the forge. Ironmaking was not his only interest, government was another. In 1848 he was elected state senator and became the one person most responsible for bringing the New Jersey Midland Railway to Bloomingdale after many years of effort. He was a respected citizen as well as an expert ironmaster.

Today a supermarket, other stores, and a paved parking lot cover most of the Bloomingdale forge site. Even the raceway has been obliterated. Besides a few fragments of slag, all that remains is the little forge building, hard pressed between modern buildings. It has been renovated to some extent, but the original walls are intact. Until a few years ago, the name Ryerson was still on the door.

CLINTON FURNACE

The partially restored ruins of the Clinton furnace stand four miles north-northwest of Charlotteburg and about a mile-and-a-half north of the Pequannock River. The original construction of the ironworks was started here July 15, 1826, by William Jackson of Rockaway, New Jersey, who had come to the Clinton region a few years earlier in search of a site for an ironworks.[1] He not only found an ideal location with the necessary waterpower provided by Buck and Cedar Ponds but also some of the most beautiful natural scenery in all New Jersey. Jackson purchased a tract of

Clinton Furnace. *Photo by Vernon Royle, 1905.*

nearly 1,000 acres, including three principal waterfalls. It was the answer to his dream. Here he built a furnace, forge, sawmill, and gristmill.

William's father, Stephen Jackson, was the proprietor of the old upper forge at Rockaway. In 1812 the father turned it over to William and another son, John D. Jackson. After a short time, the two brothers sold the forge to a third brother, Colonel Joseph Jackson, who since 1809 had been the owner of the nearby lower forge. In 1820, William and Joseph brought out "the first bar of round and square iron ever rolled in this country" at Samuel and Roswell Colt's rented rolling mill in Paterson. Elated with their success, the two brothers entered into an agreement on January 26, 1822 to construct a mill of their own at Rockaway. By November of the same year their own rolling mill was completed. William sold his half interest in the plant to his partner in 1826, putting the money into his new venture at Clinton.

Much of the subsequent history of the Clinton ironworks can be found in the writings of J. Percy Crayon who once owned the furnace ledgers and records. For more than forty years he was familiar with this region, having first visited the spot in the early 1850's and having resided there for almost twenty years before moving about 1890 to a locality now known as Denville. A photographer, writer, and school teacher, he was also a veteran of the Civil War. Through articles he wrote which appeared in the *Warwick Advertiser* in June of 1888, and in the *Evergreen News* about 1902, the history of the ironworks has been preserved, but the ledgers and other records have disappeared.

When Jackson began the furnace at Clinton in 1826, he also undertook the building of a "sawmill (at Falls No. 1) and forge (at Falls No. 3)" as well as houses and roads in the area. A store was also erected where his brother-in-law, Silas D. Halsey, acted as clerk and bookkeeper for the works. Also associated with Jackson were his two sons-in-law, Freeman Wood, who later became a judge, and John F. Winslow, who was said to be the builder and later an "owner" of the famous Civil War ironclad, the *Monitor*.[2]

Jackson soon found that the construction of his ironworks at Clinton was costing considerably more than he had anticipated.[3] As a result he lost control of the property to James Wheeler, who in turn soon sold the works to John F. Winslow and Freeman Wood.

Taking possession May 7, 1833, the new firm completed the furnace and gave their enterprise the name of the Clinton Ironworks.

The furnace, under the care of Bartholmew Goble, was fired for the first time September 17, 1833. The first casting of pig iron was made October 3, and the furnace continued until February 5, 1834, when it halted for necessary repairs; the second blast was started with the furnace being fired on April 26, 1834. Ten days later the first casting was made. On July 27, the furnace once again needed repairs and operations were resumed September 1. The continuation of the second blast lasted for seven months. Castings, made about every 17 hours, usually produced from one to two tons of pig iron, plus other forms of iron.

After the second blast repairs and remodeling followed. The third blast began on July 15, 1835. A little over 2,000 bushels of charcoal were used in the furnace. The first casting was on August 10 and after 30 castings the furnace was again cooled for repairs. This time the furnace lining had to be replaced at a cost of $340. These frequent interruptions must have been very irritating, for it was not uncommon for furnaces to have blasts that ran for a year or more.

The newly lined furnace was fired again on October 20, 1835, and evidently it worked well for by January 11, 1836, it had made 148 castings of about 12 hours each, averaging nearly two tons of pig iron for each casting. During this blast a new owner, named Pratt, acquired the ironworks on May 1, 1836 and the firm's name became the Clinton Manufacturing Company. After a number of experiments to improve the quality and the production of the iron, Pratt abandoned the furnace in 1837 and soon erected a forge in the vicinity.

According to Crayon, the reason for the closing of the Clinton furnace was the lack of charcoal supply—the timber having been exhausted. In view of the short period of time that the furnace was active, it seems rather unusual that this could have happened. If it is true, it indicates poor planning for one of the requisites for an ironworks is to have a large enough tract to take care of all charcoal needs for many years to come. If the proprietors at Clinton, in four years time, used up all their timber, then the tract was much too small to service both a furnace and a sawmill. It is of more than passing interest to note that in 1837, when the furnace closed down, iron prices were at the highest they had been since 1825 and they did not decline appreciably until the early 1840's. Closing down is hardly a normal move for a proprietor of a furnace to make when prices are at their peak.

While the furnace and sawmill were working, a number of men were employed to haul to market the iron and lumber cut on the property. These trips took them to Rockaway, New Jersey, where some of the iron was used at the rolling mill belonging to Colonel Joseph Jackson, as well as to Dover, Newark, and New York. On the return trips they brought back ore, machinery, and provisions to Clinton. The ore is said to have come from mines at Ringwood, Hibernia, Mt. Pleasant, Ogdensburg, nearby Uttertown, and Hamburg—an impressive list for a furnace operating as short a time as Clinton.

The forge that Pratt erected, after abandoning the furnace, consisted of three fires and used the same waterpower. Peter Brown, the father of John P. Brown of Newfoundland, New Jersey, became the general manager of the works for Pratt and ran the forge until 1850. Under the separate management of John McAlvannah, the anchory fashioned iron into ship's anchors.

The forge and properties were taken over by the Maryland Company in 1850 and managed by John Raymond and his son John, Jr. The Maryland Company abandoned all ironmaking at Clinton in 1852. This was probably due to the low price of iron which in 1851, had reached the lowest point in 26 years and what also proved to be the lowest price for the next 29 years. In 1847 the price of iron ranged from $70 to $77.50 per ton. By 1851 it had dropped to about half, fluctuating from $33.50 to $41.00 a ton. At this time imported iron was available in New Jersey at a lower price than it cost local ironmasters to make themselves. It was not a healthy situation and many other ironworks besides Clinton were abandoned in the 1850's.

When Crayon first saw Clinton in the middle of the nineteenth century, many of the buildings were still standing. The houses were occupied and a double log house served as a school as well as a church. The framework of the old forge and anchor shop, and the bellows house, were still standing. Machine parts were also scattered around. At that time the old furnace had 10 or 12 feet of brick work on top which was later removed, yet by 1900 it still looked much the same. Among other things, Crayon noted that the persons named in the old ledgers balanced their accounts before moving on to other places.

In October, 1859, J. W. Orr of New York City and his brother visited the Clinton ironworks site to write an illustrated article for *Harper's New Monthly Magazine.* It appeared in the issue of April, 1860, entitled *Artist Life in the Highlands.* They give a graphic description of the impressive wild scenery and tell of standing on the old flume that ran to the furnace. From the flume they could see the remains of the old 50-foot high wooden dam that ran between two perpendicular rock ledges at the Upper Clinton Falls. Among the illustrations in the article is an engraving of these falls as well as one of the Lower Falls, another source of waterpower for part of the works.

Later, the East Jersey Water Company purchased the property. On September 24, 1900, the Clinton furnace site with its ruins became part of the Pequannock Watershed Supply System owned by the city of Newark. Construction of a large dam has changed some of the landscape, but the view of the falls, the furnace, and the natural beauty of the area is well worth a trip. The furnace is one of the best preserved ones in the New York-New Jersey area.

FREEDOM FURNACE AND FORGE

This hot-blast charcoal furnace was erected on the Ringwood River, midway between Pompton and Ringwood in 1838, by Peter M. Ryerson, whose father, Martin, owned the Ringwood, Pompton, and Long Pond ironworks. Peter, born in Pompton on June 20, 1798, had helped his father with the management of the various works for a number of years and was now experienced in the art of iron-making. Through his efforts, the Morris Canal Feeder was built, starting at the Pompton works. This proved to be a big contribution to the success of their ironworks, as the canal provided a means of transportation for their iron much less expensive than previous ones.

Freedom Furnace. *Photo by Vernon Royle, 1903.*

The Freedom furnace was also known as the "Whynockie" or "Ryerson's." Constructed largely of gneiss rock, with sandstone arches and a firebrick lining, its base was 30 feet square; its height 55 feet.[1] The name "Peter M. Ryerson" and the date "1838" were inscribed on the lintels in the furnace. These iron beams, which resemble a railroad tie in shape and size, are now on the grounds at Ringwood Manor.

In 1854 while under lease by Wallace and Concklin, the furnace made 2,000 tons of iron for the Pompton rolling mill out of ore from Iron Hill (Blue Mine), several miles to the south of Ringwood. It went out of blast in the summer of 1855. The old furnace stack stood until 1928, when it was demolished to make way for the Wanaque Reservoir, the body of water that now covers the site at Midvale, New Jersey. The stone from the furnace went into the construction of the Midvale dam, one of the many built for the huge reservoir.

Adjacent to the furnace was Freedom forge, owned by Martin Ryerson. It was equipped with 6 trip-hammers with a daily output of close to 4 tons. There were also 3 grist mills, a sawmill, a lime kiln, and 2 stores located nearby. In 1822, a surveyor's report refers to the "Winockey Forge"—an earlier name.

When Martin Ryerson made his will in 1833, he not only bequeathed to Peter his "Whynockie Forge" and "Pompton Furnace" lots, with the ironworks thereon, but also gave him the right to take any quantity of ore he wanted at any time, without any charge, from the "Ringwood Great Furnace Tract." After his father's death in 1839, the following inventory was taken of the assets at Freedom forge:

1650 lbs or iron	@$75.00 per ton	$ 61.88
tongs, bars, scalebeam and other implements		60.00
24,000 bushels of coal		1200.00
2 mules, Tobe and Prinny		40.00

Peter, now running the Freedom forge and furnace, as well as the works at Pompton, was not as successful an ironmaster as his father and before long went into debt and lost the works during the 1840's. He did, however, achieve temporary success in another enterprise, for he became a large stockholder and a director of

Freedom Furnace. *Photo by Vernon Royle, circa 1870.*

the Morris Canal Company and for several years acted as superintendent for the company. He later became a major in the Union Army and was killed in action in 1862, the first New Jersey field officer to fall in the war.[2]

POMPTON IRONWORKS

In 1694, Captain Arent Schuyler of New York City was employed by Governor Benjamin Fletcher of New York to confer with the Indians at Minisink on official business.[1] Although only thirty-two at the time, young Schuyler was given the assignment because of his friendliness with the Indians and his knowledge of their customs and language. As he passed through the Pompton area he must have been impressed with the desirability of owning property there, for on June 6, 1695 he purchased 5,500 acres of land on the Pequannock and Pompton Creeks from the Indian owners for "wampum and other goods and Merchandise to the value of two hundred and fifty pounds current money of New York." [2]

On November 11, 1695, Schuyler and Major Anthony Brockholst, also of New York, acquired the title to the same land from the East Jersey Proprietors, part of which was for mining.[3] The northern portion of the 5,500-acre tract, which consisted of 1,250 acres, was called the Pompton Patent. It was at Pompton that both Schuyler and Brockholst settled in 1697. Arent Schuyler moved to New Barbadoes Neck in 1710 but his son Philip stayed on at Pompton. Whether either had a hand in erecting the first ironworks at Pompton is not known, nor is the date of construction. A survey of 1726 mentions an ironworks "now building" at Pompton, but gives no other information about it. Presumably it was a bloomery.

At a later date production was greatly increased by the erection of a furnace. Tradition has it that the Pompton works supplied ball and shot for the French and Indian Wars. It also supplied the usual firebacks and utensils for local residents, and pig and bar iron for the trade.

On December 15, 1774, Casparus Schuyler, the eldest son of Philip, sold the Pompton ironworks to Gabriel Ogden who, with

Pompton Furnace. July 13, 1854. *Courtesy of Mrs. Barbara Lang.*

his family, already had an interest in the Vesuvius furnace at Newark.

As early as the winter of 1776–77, the Pompton works was busily engaged in filling orders from General Henry Knox for over 7,000 cannonballs from 4 to 18 pounds in weight, as well as for 10 tons of grape shot. On March 12, 1777, Washington wrote to Major General Philip Schuyler, from Morristown: "I have given directions to the Managers of the Iron works [Gabriel Ogden of the Pompton furnace and Charles Hoff Jr. of the Hibernia Furnace] to have the shot for which you inclosed General Knox's orders, conveyed to some convenient landing place, 'till the river opens, and then be forwarded to you with all dispatch." [4]

Four days later General Knox was at Pompton inspecting the ammunition Ogden had made. He then sent it to King's Ferry for transportation by water to General Schuyler at Albany to reach him as soon as possible "as they [the shot] are much wanted at Ticonderoga." Knox also gave instructions to "reserve as many sorts as will be wanted for the forts in the highlands."

Like so many of the other ironworks that were casting supplies for the army during the war, Ogden's Pompton works was soon hampered by the lack of manpower. He evidently applied to Washington for more men to keep the works going and thus fill the urgent orders for ammunition. His request was granted by a letter written on May 2, 1777, at Morristown, by Tench Tilghman (one of Washington's aides) to Brig. General Nathaniel Heard who was at the time stationed at Pompton: "Mr. Gabriel Ogden of Pompton is employed to cast Cannon Ball and Grape Shot for the public use; but from the great Scarcity of hands he is not able to go on so briskly, as the Service requires. His Excellency therefore desires, that you would permit him to employ about forty of your Men upon such terms as he and they can agree. As the Works are at Pompton, these Men can, upon any alarm, take up their arms, and be useful as soldiers." [5]

The papers of Major Sebastian Bauman of the artillery, who was under General Knox, show the activity at the Pompton ironworks, as well as the problems of production that Gabriel Ogden had to face in the fall of 1779. Bauman seems to have been a frequent visitor to the Pompton works in carrying out his superior's orders, especially during this period. On October 4, 1779 he wrote from Pompton to Brig. Major Shaw:

> I received your letter of the 29th this day at my arrival at Pompton. You'l please to acquaint the General that no shells have been cast here during my absence. Nothing but carcasses of which you acknowledge to have received the profile, and which you say is so far approved except the equality or equal roundness of the metal which was done so by my order and direction, for reason of holding the more composition than otherwise they would and likewise for the saving of Iron, which I conceive to be of no manner of use at the bottom. Because when the carcass is filled, it will way more than a life shell, and I still believe them better with less Iron had you not suggested to me that the general may have reason to have them otherwise. I have therefore directed Mr. Ogden to make them some what thicker at the bottom.
>
> I have ordered three men to Mount Hope. There will be a number of shells in a few days which I intend to get ready in case any should be called for. Mr. Faesch has much solicit me to return to Mount Hope on account of the Pattern, which was made in a most stuplified man-

ner. Mr. Ogden has cast an Iron mould for the casting of shells, but made it too large, which is therefore entirely useless. But he intends to try again, and if that wont do, he will try no more. I fear they will cast but few shells here during the present Blast. I have nothing further to add, but that I am in hopes to see you soon. For I am exceedingly tired by too much riding and I and my old horse begin to feel the fatigue of it . . .

Bauman was not soon to be relieved, for the following day, General Knox wrote to him from West Point regarding very urgent orders for more shells to be cast at the Pompton furnace, as well as at Mt. Hope:

I shall want a great number of shells proper for the French 9 inch Mortars, and have given directions to Mr. Gabriel Ogden of Pompton to have *one thousand* cast at his furnace. But I am afraid that he is not as fully acquainted with the nicety of the business as may be requisite. Therefore beg the favor of you to go to his furnace and prompt him to make the utmost dispatch in his preparation for casting the shells, and when he has cast some of them I wish you to examine and prove them critically and fully, so that I may be able to judge of the certainty of my supplies. The exigence of the service requires that this business be executed with the greatest dispatch, and I have the fullest confidence that you will expedite it to the utmost of your power.—Mr. Ogden may also cast *two thousand* 18 pound balls—but in such a manner as not to retard casting the shells a moment. Immediately on your arrival at Pompton you will please to inform me whether the shells can be cast—if so, when they can begin, and how many they can cast in one day—

Mr. Faesh at Mount Hope informed me that his furnace would soon be in blast. You will please to ride there and see when, and enquire whether he can cast shells, and when he can begin—how many in a day—and the price,—of all which you will inform me as *soon as possible*.

I have not mentioned the number of shells that will be wanted, as they will greatly exceed anything that can be done at one furnace in any reasonable time—therefore I shall employ as many different furnaces as will best answer the end.

Bauman had left Pompton before the above letter from General Knox reached him. However, he was back in Pompton on October 13 when he wrote to Knox:

I came to this place again this day agreeable to what I wrote to you in two former letters, which I hope you have received. But there has been no shells cast as yet, occasioned by some existence in the furnace, but assured me, which I am to inform you that without fail some shells will be cast by Friday or Saturday and ready for prove by Sunday. Mr. Ogden has cast some 18 pound Ball, which I can assure you are good and must approve of them.

I met Mr. Faesh at this place, as I informed you in my letter of the 11th instant, that he would meet me here, in order to lay before you through this, that his furnace will be in Blast the 23rd and can begin to cast shells the 25th and one third more in a day than Mr. Ogden and that his furnace will be in Blast about Eight weeks and during that time could cast 1500 shells. As for the price he cannot fix untill he has made a tryal.

I am likewise to inform you from both the gentlemen, that it would be requesit to apply to his Excellency for his application to the governor, letting him know, that while these furnaces are employed for the service, should the militia be called out, that the workmen employed therein might be exempted, as long as they are so employed.

The sooner an answer with regard to Mr. Faesh, the sooner will he forward the business in hand.—I would likewise wish to let me know, or these gentlemen how many shells of each sort they are to cast in order that they might without any interruption go on with the matter in the best manner possible.

S. Bauman

P.S. I must again request to let me know by the first opportunity concerning the diameter of the French 9 inch shell as I wrote before the profile which Mr. Ogden has from Colonel Hopkins is 9.6 Inches and those I took with [me] are but 9.5.—

Five days later, on October 18, 1779, Bauman again wrote to Knox, but this time with more reassuring news from Pompton:

I have the pleasure to inform you that Mr. Ogden cast some shells yesterday, which I this day proved, and out of ten[,] seven proved good. From the success Mr. Ogden had in the first casting, think to go on with hopes of still greater success. So that in a few days be able to judge by what time to deliver the number of shells required.

I should have answered your letter of the 11th before this but waiting until I could give you an account how those shells cast yesterday turned out, which I think are tolerably good. Mr. Ogden will go about immediately to cast those carcasses you desire.—

I want your order, with regard to Mr. Faesh who as I wrote before says that his furnaces would be in blast by the 23, and commence to cast by the 28th therefore waits your order for what number he should cast if time and what size. But I fear he wont be able to cast any before the first of next month for I find he has to make his moulds first, which takes a deal of time, which I fined so at Pompton as they have no other mode for the casting of them but in sand which is both troublesome and tedious.

Mr. Ogden casts very fine 18 pound shot and the shells I think will do, considering the manner in which they are cast. Those men which are sent here, I shall keep and employ in the manner all ready mentioned untill you'd please to order them otherwise.

I shall find but little to do here in a few days after they are going on in a regular manner to work. Would be glad to know your order concerning me, nothing shall give me more pleasure than to order me to whatsoever I may be servicible, both to you or the publick.

In answer to Major Bauman's letter Knox wrote from his headquarters at New Windsor on October 21, 1779: "I received your favor of the 18th instant—I am happy to find out Mr. Ogden has success in casting—I wrote you by Mr. Hodgdon how many shells that I would have cast at Pompton and at Mt. Hope by Mr. Faesh. They both must be informed minutely that they will be paid only for the Shells which shall stand proof at Furnaces. As soon as you are assured that Mr. Faesh can cast the number of Shells required you will please to return to camp."

Major Bauman's anticipated departure from Pompton to return to camp was soon postponed, for Gabriel Ogden was starting to experience real difficulty in casting the French shells. On October 22, 1779, Bauman wrote to Knox to tell him of the predicament at the Pompton furnace:

I stated to Mr. Ogden part of the contents of your letter of the 17th Inst. which I received by Mr. Hodgdon and to which Mr. Ogden made no objection, with Regard to the price, only, that the Board of War he hopes would not let him be a loser by it.

I wrote of Mr. Faesh two days ago, including to him that very same paragraph. . . . You will please to remark, that in my letter to you of the 18th Inst. I acquainted you with the success Mr. Ogden had in the first casting. Which was really so for two days successfully, and from which, as I wrote, he would by the middle of the week be able to form an estimate by what time he could deliver the number of Shells required. But I am sorry to inform you, that after the two first castings all the shells turned out bad. There is fifty cast, twenty of which are good, fifteen may do with some mending; and fifteen are entirely Bad, and useless, the reason I cannot see into; and it has Discouraged Mr. Ogden so, that he has stopt casting any more until Monday when he intends to begin again in another mode, by which he promises himself better success.

I find his hands are new and few, all the Shells cast are good at the bottom, but about the fuse hole and ears they look Very bad, caused from the dross of the Iron, which it seems always gathers there—and which causes them not to hold wind. Which fault however Mr. Ogden thinks to prevent by some new contrivance.

I shall send a Carcass and a Shell by next Monday to your quarters, and I would willing wish that in the meantime a few pounds of powder and some fuses might be send here in order to burst one of those shells, for reasons they being made upon an entire new construction from the rest, therefore would willingly know in how many pieces they would Break.

I received your favour of the 21 Inst. this evening. I had before suggested to those gentlemen that they could only expect to be paid for those shells which were good and stand prove before they are carried from the furnace wherein they are cast.

The tribulations of running a furnace during the war were becoming more and more apparent to Ogden. Not only was he evidently working overtime to fill the Continental Army's rush orders for cannonballs, shells, and shot—with mixed results—but he was confronted with the lack of experienced workmen as well as an overall shortage of manpower to run the ironworks.

A continuing demand for ammunition kept Ogden engaged in producing it at his Pompton furnace as late as the summer of 1780:

[George Washington to the Board of War, Hdqtrs. Col. Deys, Bergen County, July 8, 1780]

Inclosed is a letter from Brigr. Gen. Knox on the subject of an instant provision of shot and shells, and proposing the employing of Faesh's and Ogden's furnaces for this purpose. As the matter is of the utmost importance, and requires an immediate decision, I intreat it of the Board. And should no arrangements have been made on this head I think these furnaces would answer the intention, and should be engaged in the business without a moments delay.[6]

[General John Lamb to General Knox, Ramapaugh, June 29, 1780]

We have at Pompton and Mt. Hope Furnaces between five and six thousand 18 pound balls and three thousand shells, for the French 9 inch mortars, but I have not been able to have them transported to West Point, by reason of the utter inability of the Quarter-Master General's department.[7]

The importance of the Pompton ironworks to the American cause can be readily seen from these letters. Pompton, itself, was important because of its strategic location. It commanded the approaches from the Hudson River—either by Smith's Clove and the Ramapo Valley, by Ringwood, or by Paramus—to Morristown, and fortifications were erected at Pompton in 1777 by the New Jersey militia under General Nathaniel Heard.

Besides Gabriel Ogden, who ran the works, Uzal, Charles, and Moses Ogden, as well as Peter Schuyler, all seem to have had a financial interest in the Pompton enterprise. Charles Ogden acquired Uzal's share in November of 1788, and Peter Schuyler's interest in March of 1797. He then sold all of his holdings at Pompton to Martin Ryerson by deed dated April 18, 1797, which included "all buildings, furnaces, forges, sawmills, grist mills, out houses, waters, streams, water courses" as well as "cordwood set in pitts . . . patterns, flasks and tools belonging to the furnace at Pompton." Moses Ogden sold his interest to Ryerson also by deed of the same date, thus making Martin Ryerson the sole owner of the Pompton ironworks property.

The new owner had been born in 1751. He became an able ironmaster and expanded his holdings by purchasing land at Bloomingdale and Whynockie where he erected forges. His final effort to establish an iron empire of his own was the acquisition of the Ringwood and Long Pond ironworks, as well as half of the Charlotteburg property. Despite the number of enterprises (including Freedom forge), Ryerson ran his works in an efficient and profitable manner. He was accused of selling ammunition to the British during the War of 1812, but cleared himself of this charge. He was in fact making shot for the American forces. A notice reading "Headquarters Q.M.D. Sept. 9, 1814" states that "Such persons as are employed by Martin Ryerson in making shot, will continue in that employ until further orders."

An inventory taken of Martin Ryerson's holdings in 1840, a year after his death, lists the following assets at Pompton furnace:

11.14.3.20	bar iron at the landing	@$75. per ton	$ 881.08
1,015	tons of ore at the furnace	@2.50 per ton	2,537.50
106,879	bushels of coal on the furnace bank	@.05 per bu.	5,343.95
20	wheel barrows in the different jobs		20.00
	old furnace patterns in carriage house		15.00
1	ton of bar iron in store cellar		75.00
	ruffage bar iron in store cellar		30.00
3	ore boxes and 2 stone frames		18.00
5	old cast iron wheels		5.00
3	yoke of oxen at furnace		205.00
2	old waggons at furnace		15.00
1	mule, Blackbird		40.00
8	picks ($4.00) 2 ox carts ($5.00)		9.00
24	shovels		9.00
2	scows ($6.00) 12 crow bars ($9.00)		15.00
1	large gudgeon and hammer faces		39.50
350	bushels of potatoes in furnace lot	@.25 per bu.	87.50

After Martin's death in 1839, his son, Peter M. Ryerson, inherited the "Pompton Furnace lot" and the works thereon. Peter had built a hot-blast charcoal furnace in 1837 on his father's property in Pompton. Close by he had erected a rolling mill in 1838. Both of these were constructed shortly after he had completed arrangements for the Pompton Feeder, a branch of the Morris Canal, which would make it possible for the canal boats to dock right at the ironworks. By doing this he had high hopes for the success of his new enterprises, but unfortunately just as he was about to profit from his investment, reduction of the tariff put him in financial difficulty. He continued to struggle to make the works profitable, but was later forced to sell his property to satisfy his creditors. As early as 1842, the sheriff advertised the following for sale:

> One lot known as the Furnace Lot, said to contain 222 acres together with all the improvements thereon erected, such as water power, mill

> dam, docks, raceways, rolling mill, dwelling houses & etc . . . the adjacent Post farm lot of 162 acres, one known as the Van Houten farm of 152 acres and the Johnson lot of 150 acres.

In addition to this property at Pompton, other tracts owned by the Ryersons at Ringwood, Long Pond, and Whynockie were also put up for sale, but never actually auctioned off. Throughout the next eight years the properties were often listed for sale by the sheriff with advertisements circulated as late as August 1850. But the sales were never held. In the interim, the rolling mill had been enlarged in 1844 by a Horace Grey, who must have purchased or leased it from the Ryersons.[8]

The Pompton ironworks property seems to have been eventually sold in sections, for in 1854 the furnace was owned by William C. Vreeland and his associates of Bergen Point, New Jersey.[9] In the same year, James Horner acquired part of the property and began to turn out file steel and hand-made files. He employed a young salesman, James Ludlum, son of General Ludlum. In 1863 a steel works was erected to produce railway-car springs and cast steel.[10] By 1864, Horner had taken Ludlum into partnership and the firm was then known as Horner & Ludlum. In 1875 the business was called the Pompton Iron and Steel Company, with Ludlum the president. He died in 1892 and was succeeded by William E. Ludlum. In 1898 the company became the Ludlum Steel & Spring Company. Nine years later it was moved from Pompton to Watervliet, New York, thus bringing to a close over 175 years of iron manufacturing at Pompton.

At eighty-six, Charles Edward Turse, a resident of Pompton Lakes, vividly recalled working in the Ludlum's spring and file shops as a boy in the early 1890's. The work was so hot, hard, and heavy for only a dollar a day that he gave it up after a couple of years. The workers made several kinds of springs: coil, flat, engine, and springs for locomotive tenders, most of which went to the old Pierson works at Ramapo, New York, he said. A 40-foot overshot waterwheel and a turbine powered the rolling mill and the steel works until the Pompton dam was washed out in the flood of 1903. The present Furnace Inn, on Hamburg Turnpike at the corner of Hemlock Road, was once the storehouse for the files. Here in this building they were packaged and shipped to all parts of the globe.

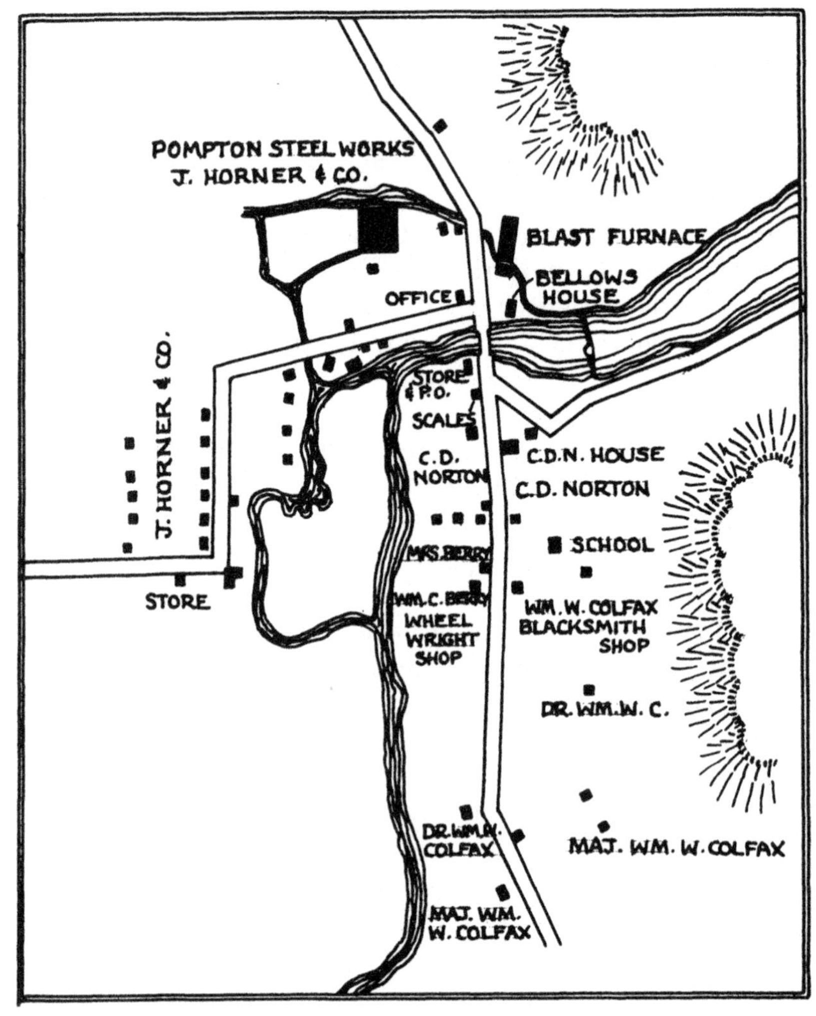

Pompton. From Corey's wall map, Philadelphia, 1861.

In an interview with the author, Mr. Turse insisted proudly that "the finest tool steel in the country was made at the Ludlum works in Pompton."

Behind the Furnace Inn stands an old stone pillar that once supported the charging bridge from the hillside to the top of Peter Ryerson's 1837 furnace. It is located about sixty yards east of Hamburg Turnpike on the north side of Hemlock Road. The furnace itself is completely gone. The river and falls that supplied the power are on the south side of the road. This site is an excellent natural location for a furnace and one would immediately think the original furnace was also located at this ideal spot. However, a map of General William Colfax's farm dated June 1, 1822, made by Abram Ryerson, Jr., clearly shows a furnace across the river and downstream a short distance on the west side of the Hamburg Turnpike, almost opposite the present Grand Union store. Ryerson not only marked it "Furnace," but drew a picture of it and its waterwheel. It is also interesting to note that William Nelson in 1901 wrote that Pompton was "important on account of the iron furnaces and forges and the shops where cannon balls were made for the American army, in a long, low building almost directly opposite the present Norton House." [11] This house stood on the present Grand Union property. Mr. Turse remembered his father's saying that there used to be a large slag dump located on the present animal hospital property on the west side of the Hamburg Turnpike.[12] Erskine's 1777 map of New Jersey shows a manufacturing site on the north as well as south side of the river here. He made another map of the Pompton region in 1778–79, with a symbol for a single works on the north side and for two on the south side, designated "Pompton I. W." [13] The question is, was the furnace shown on the south side of the stream on the 1822 map, the original one or not? From the present contour of the land it does not seem to be the preferred site, but contours can change over the years.

The thought arises that there may have been three furnaces at Pompton, rather than just two. The original one may have been erected on the same site that Peter Ryerson chose for the third one. The second furnace then would be the one shown on the 1822 map on the opposite side of the river. Perhaps some day data will come to light which will give the answer.

The Pompton Women's Club has placed a marker at the furnace site on Hemlock Road which reads:

OLD POMPTON FURNACE
BUILT CIRCA 1700
ARTICLES FORGED FOR ARMIES
OF FRENCH AND INDIAN WARS

"Ironworks" for "furnace," would be more appropriate as there is nothing to substantiate a furnace at Pompton even as late as 1750, let alone circa 1700. It is very likely that the original works consisted of a bloomery only, where iron was made directly from the ore.

A few firebacks made years ago at the Pompton works are still in existence, prized by their owners. The Paramus Historical and Preservation Society has one. It bears the words "Pompton Furnace" plus a decoration consisting of a ribbon scroll and a basket of fruit, a reminder of the days of this old ironworks.[14]

WAWAYANDA FURNACE

The well-preserved ruins of the Wawayanda charcoal blast furnace may be found at the outlet of Lake Wawayanda, formerly known as Double Pond, in Sussex County, New Jersey. This stack, although located about ten miles west of Sterling Lake and the Ramapos, is included to preserve some of its history.

The Wawayanda furnace was owned by Oliver Ames and his three sons, Oliver, Jr., William, and Oakes, who also had shovel factories and forges in Massachusetts at Braintree, Canton, Easton, and West Bridgewater. Oliver, Sr., born April 11, 1779, served his apprenticeship under his father and before he was twenty-five was running the factory. As the business grew and expanded under his leadership, Ames' shovels soon became a well-known and leading article in the nation's hardware trade and remained so for many years because they were lighter in weight than the heavy implements of the time and, though less durable, were easier to handle.

In 1845, Ames and his sons decided to expand their holdings to New Jersey and build a blast furnace at Lake Wawayanda. William, who acted as a roving engineer for their various plants, supervised the construction of the Wawayanda furnace, which was begun in 1845 and completed in 1846. The initials "W.L.A." and the date "1846" still stand out clearly on one of the lintels in the main arch of the stack. The iron produced at this furnace was particularly good for making wheels for railroad cars. In 1854, under the supervision of J. A. Brown, who managed the ironworks, 1,852 tons

Wawayanda Furnace. *Courtesy of Fred Ferber.*

of iron were made for this purpose during a 40-week period.[1] Wawayanda furnace was arranged so it could work on either a hot or cold blast, but the New Jersey State geologist, William Kitchell, reported that the cold blast only was used when he visited there in 1856, as "a superior article of iron was thus produced."

The Wawayanda furnace was 11 feet wide and 42 feet high inside. To produce a ton of iron the following materials were used in 1856: an average of 180 bushels of charcoal, 2 tons of ore, and ¼ ton of limestone. These were introduced into the furnace at the top with 50–60 charges during a 24-hour period. Each charge would usually be made up of 11 baskets of coal (22 bushels), 525 pounds of ore that had been reduced to nuggets about the size of hickory nuts, plus an average of 40 pounds of limestone, the quantity depending on the quality. These charges would yield from 7–8 tons of iron a day, with the casting being done at noon and midnight.

Most of the ore came from the Wawayanda mine, located 2½-miles northeast of Wawayanda Lake and ½ mile from the New

Wawayanda Furnace lintel.

York state line. This mine consisted of five separate openings, the largest having been worked 110 feet in depth; and 100 feet along the vein which was 12–14 feet in width at the bottom of the shaft. This was the most southeasterly of all the openings. A tract of 6,000 acres supplied the charcoal.

At its heyday, the furnace community of Wawayanda consisted of a post office, store, gristmill, and sawmill as well as the homes of the workmen and the other buildings usually connected with an ironworks.[2] Today only the furnace and a mule barn remain of what once was a busy little village. The furnace was not used very much for it was operated on and off during a period of only ten years, having its final blast in 1856. During the early 1850's it was out of blast for about two years.

As for the Ames family, Oliver, Sr., died in 1863 when Oakes and Oliver, Jr., took over the management of the shovel factories and other related ironworks. They increased the value from $200,000 in 1844 to $4,000,000–$8,000,000 during the Civil War period when they filled government orders for shovels and swords. Oakes Ames went on into politics and was elected to the Massachusetts House of Representatives, and it has been said that later he was more responsible than any other individual for the construction of the Union Pacific Railroad.

Today's plans for the Wawayanda tract are promising and will appeal to varied tastes. Fred Ferber, who prefers acquiring land in its natural beauty rather than paintings and sculpture, has bought 2,800 of these acres; New Jersey, under its Green Acres program, has taken over the remainder, which includes the lake and furnace sites.

Peter E. Scovern, whose initials are his by-line, reports on one rare day in June of 1963, spent hiking in this "Unspoiled Wilderness" as he titled his column:

> There's more to the out-of-doors than just hunting and fishing. A sportsman can get as much enjoyment out of watching the antics of a half-dozen grouse chicks in the spring as he gets from bagging a brace of the mountain birds in the fall. And to some, conservation and an appreciation of nature are all that is needed for a full life in the open.
>
> Last Sunday, we spent most of the day with Fred Ferber, owner of Sussex Woodlands. . . . Along with a party of about 20, we spent

the entire day hiking to various spots in the Sussex Woodland holdings.

There are about 7,000 acres in the entire Wawayanda tract. . . . Ferber took over almost 3,000 acres, most of them in West Milford . . . the State [has taken over] the Sussex County acreage.

Ferber and the State have different ideas about what should be done with the unspoiled mountain wilderness at Wawayanda. The State, according to the announced plans of the Conservation Department, would establish a new State park, with campsites, picnic areas, playgrounds, etc., and honeycomb the mountains with roads necessary to take care of the 5,000 or 10,000 visitors expected daily at the park.

Ferber, who will proceed with his program on the acreage he presently owns . . . says his aim is to retain the pristine beauty of the mountains. Instead of building roads for visitors to drive to picnic areas and campsites, he wants to mark hiking trails. Enjoyment will come from foot-by-foot appreciation of the woodlands and not from a hurried glimpse out of a car window.

We hiked up the Bearfort Mountain last Sunday. It was a rugged climb over rocky ridges and through seeming miles of flowering mountain laurel. At the top of the peak is Terrace Pond, a natural body of water surrounded by perpendicular walls of rock. It is a spot of unparalleled beauty, and its esthetic value is most apparent when a person has to climb a mountain to see it.

During the day's excursions into various scenic spots, I watched a flock of mallard ducklings crossing the edge of one pond like a flotilla. I kicked up a covey of baby grouse, with Mamma also on hand to distract and decoy the intruder. I saw a flashing mountain stream filled with native trout. I observed wood ducks nesting. And more. And everywhere there was natural beauty of the woods and waters.

Of course, there are discordant notes even in paradise. Ferber's workers have had to remove tons of garbage and rubbish left by campers and picknickers. And even shimmering Terrace Pond, which can be reached only by a tortuous hike up Bearfort Mountain, has beer cans in the water. But they were noticeable because they were so few.

New Jersey needs every acre of state park and forest it can lay its collective hands on. But more importantly, the Garden State needs the primitive tracts that can come only when an individual like Ferber initiates the action and carries it through.[3]

II

FORGES AND FURNACES IN SOUTHEASTERN NEW YORK

Augusta Forge in ruins. Elizabeth Oakes Smith, *The Salamander,* 1848, frontispiece.

AUGUSTA WORKS

In 1783 Peter Townsend, owner of the Sterling ironworks, sold about 6,000 acres—called the Augusta Tract—to a cousin who was also his son-in-law, Captain Solomon Townsend. The new owner built a forge there that same year.

Solomon Townsend was born at Oyster Bay, Long Island, in 1746, and his love of the sea was evidenced early. When only twenty, he was made captain of a brig owned by his father. He later became the master of the *Glasgow*, owned by Thomas Buchanan, a New York merchant, and he was abroad with this vessel at the outbreak of the Revolution. Eager to return home, he obtained the following oath from Benjamin Franklin, minister at the French capital:

> Passy, near Paris, June 27, 1778.
>
> I certify to whom it may concern, that Captain Solomon Townsend, of New York, mariner, hath this day appeared voluntarily before me, and taken the oath of allegiance to the United States of America, according to the resolution of congress, thereby acknowledging himself a subject of the United States.
>
> B. Franklin [1]

Soon after returning in 1778 he went to Orange County, New York. There early in 1783, he married his cousin, Anne, daughter of Peter Townsend. Shortly after, the Augusta forge was completed and later an anchory. Both were on the west bank of the Ramapo River at the foot of a waterfall about seven tenths of a mile north of the present railroad station at Tuxedo, New York. In the early

1900's, before the widening of State Highway 17 through Tuxedo, the works were described as being immediately below and opposite the great boulder resembling the bow of a warship, known as "Man-of-War Rock," which unfortunately was demolished in favor of a broader road.

At his Augusta works, Townsend produced bar iron as well as flat and square iron, crow bars, share moulds, anchors, hollow ware, and various castings. A large bar iron store in New York City served as an outlet for the products of the ironworks, which were transported by wagon and boat.

Townsend had a residence at Augusta, as well as one in the city. Although most of his time was probably spent in New York, he was often at Augusta where he divided his time between the ironworks and his vegetable and flower gardens, upon which he lavished attention. His small hand-written diaries have entries fondly describing the sizes and colors of the vegetables and flowers he grew. His hobby offered release from the vagaries of the iron business.

William Townsend, a brother of Peter, apparently joined him as a partner in the Augusta works about 1792, at which time they employed nearly 40 people.[2] In 1804 they decided to rebuild the Augusta forge and enlarge it. By October 15 the work was under way. The lower bellows wheels were removed as well as the posts and roof at the northern end and the remaining beams supported with studs. Considerable blasting was necessary to remove the native rock in order to lay the new foundation. By early December the rebuilt forge was completed and on the thirteenth Augusta began to make iron again.

The forge was an odd-shaped building. According to specifications the wall next to the river was 51 feet 8 inches in length, whereas the opposite wall, next to the hillside, was 46 feet 5 inches long. The wall on the stream side was constructed as near the river as was feasible and followed the contour of the bank, which made it longer and not parallel to the opposite wall, nor at right angles to the end walls, of course. The wall at the lower end was 38 feet 5 inches in length. The foundation was three feet thick. The overall cost of rebuilding the forge was given as 660 pounds.

To operate the forge bellows a waterwheel 16 feet in diameter was constructed, containing 48 buckets made of boards 1¼ inches thick, 15 inches wide and 4 feet long. To make, deliver, and fit the wooden shaft to the wheel, cost the Townsends 4 pounds; affixing the cams that made the bellows open and close, was another 8 pounds. The wheel's total cost came to slightly more than 25 pounds sterling.

Horatio Spafford describes the works about 1810, in his *Gazeteer of the State of New York* (1813):

> Augusta Works, in the S., near the line of Rockland County, are very extensive, and merit detailed notice. The bloomery is a stone building, 70 feet long, in which are 4 fires and 2 hammers. On the ridge is a cistern, 70 feet long, 8 feet wide, and 10 inches deep, kept full of water for extinguishing accidental fires. There is an anchor works which makes 60 tons yearly, and where anchors are made weighing 6000 pounds. A grain and sawmill belong to this establishment, which is the property of Solomon Townsend, Esq. of New York, and to which is also attached 12,000 acres of land. The bloomeries may be made to yield 200 tons of bar-iron yearly.

Two years after Solomon Townsend died in 1811, the Augusta works were shut down forever. The deserted appearance of the little village during the early 1830's made a deep impression on Frank Forester, a noted writer and sportsman, for in his book, *The Warwick Woodlands,* he gives the following description of the locality:

> We reached another hamlet far different in its aspect from the busy bustling place we had left some five miles behind. There were some twenty houses, with two large mills of solid masonry; but of these not one building was tenanted now. The roof-trees broken, the doors and shutters either torn from their hinges, or flapping wildly to and fro. The mill wheels cumbering the stream with masses of decaying timber, and the whole presenting a most desolate and mourning aspect. Its story is soon told—a speculating clever New York merchant—a water-power—a failure—and a consequent desertion of the project.[3]

While the forge was in operation, the town of Augusta Falls had a population of almost 500. Although the coming of the railroad in 1841 stirred up new activity in cutting timber for the wood burning engines, by 1858 only 40 people were living in this once thriving community.

Solomon Townsend was a man of ability in the iron business, but events beyond his control brought him disaster. The Embargo Act of 1807, proposed by President Jefferson in an effort to resist the injurious policies of France and Great Britain, forbade American vessels sailing to foreign ports. It brought financial ruin to many American merchants and ship owners, among them Captain Townsend. It is said that this act cost Townsend almost $70,000.

Two years after his death in November, 1813, Townsend's widow sold the Augusta tract to Peter Lorillard of New York. When Lorillard purchased the estate, it had 40 dwellings, 2 gristmills, the forge, mineral rights, and thick woods. Lorillard's descendents later transformed much of the area into what is today Tuxedo Park, New York. In 1880, the Tuxedo station on the Erie Railroad was called "Lorillard." It seems fitting that Townsend, a man with a great love of gardens and sweeping vistas, should have been the former owner of the beautiful Tuxedo Park.[4]

CEDAR PONDS FURNACE

The ruins of this little known furnace lie within Palisades Interstate Park on the west bank of the brook which flows out of Lake Tiorati about a half mile down-stream. The lake, formerly known as the two Cedar Ponds, from which the ironworks derived its name, is located about five miles west of Bear Mountain, New York.

In 1765, Peter Hasenclever acquired 1,000 acres in the highlands of old Orange County. This tract of land was situated in Lot No. 3 of the Great Mountain Lots, a division of the Cheesecocks Patent.[1] It included part of the ponds as well as several deposits of iron ore. Hasenclever was at first enthusiastic about this purchase: "heaps of Iron-ore lay on the surface of the earth, and there never was a finer prospect for success. . . ."

With high hopes, he began to erect his fifth American furnace, building a dam across the outlet of the Cedar Ponds, converting them into a single body of water as a large reservoir for waterpower. He also constructed log houses, cut wood for charcoal, and built a road to this area.

Unfortunately Hasenclever was forced to abandon this project after spending huge sums of money on it. He ascribed failure to the following drawbacks: "Some of the mines which produced excellent ore vanished, other mines turned out to be sulphureous, copperish, cold shear . . . so that the ore could not be made use of; these circumstances might appear incredible if the places could not be shown."

The only mine on the property that yielded ore of a profitable quality is one that still bears his name. The Hasenclever mine lies three fourths of a mile south of Lake Tiorati. Hasenclever never completed the furnace; in fact, it seems rather doubtful whether the furnace was even half-finished when he abandoned it. He was also having problems with the ironworks at Ringwood, Long Pond, and Charlotteburg at the same time. In addition, his backers were complaining about the large sums of money he was spending compared to the small income from the ironworks. Another probable reason for terminating the Cedar Ponds project was its isolated position some distance from the more centrally located works in New Jersey.

The land to the northeast, adjacent to Hasenclever's acreage at Cedar Ponds, consisted of a large tract owned by Lord Sterling. This tract, including a portion of the Cedar Ponds and the stream flowing from them, was in Great Mountain Lot No. 2. The brook provided the only logical furnace site in the Cedar Ponds area. No part of it flowed through Hasenclever's property in Great Mountain Lot No. 3, but he could easily have been under the mistaken impression that it did, since it was near. It thus appears that his furnace site was actually on Lord Sterling's property. Erskine's maps and others show the furnace site along this stream and refer to it as "Hasenclever's Intended Works," bearing out this contention.

Others with an interest in the Cedar Ponds ironworks soon followed Hasenclever, constructing ironworks of their own along Cedar Pond Brook and working the Hasenclever mine. On June 2, 1767, Lord Sterling mortgaged his tract to William Livingston. Twenty-two years later, it was assigned to Samuel Brewster, one of Rockland County's largest landholders, and a pioneer in the iron industry. Born in 1737, he came to Orange County before 1750, and died November 29, 1821. He worked the Hasenclever

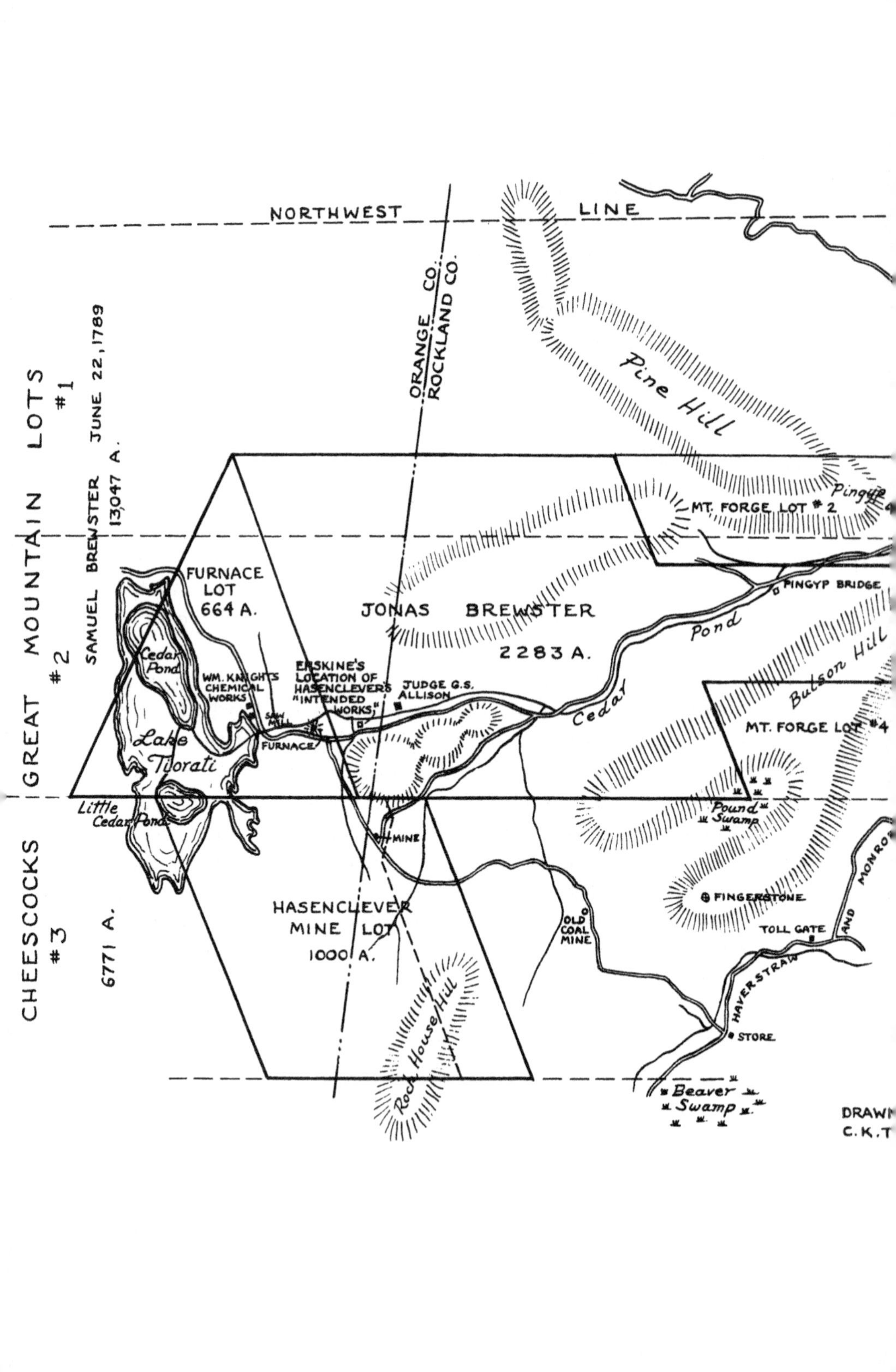

NORTHWEST LINE
ORANGE CO.
ROCKLAND CO.
GREAT MOUNTAIN LOTS
1
SAMUEL BREWSTER JUNE 22, 1789
13,047 A.
2
CHEESCOCKS
3
6771 A.
Pine Hill
Pingyp
MT. FORGE LOT # 2
FURNACE LOT
664 A.
JONAS BREWSTER
2283 A.
PINGYP BRIDGE
Pond
Cedar
Bulson Hill
Cedar Pond
WM. KNIGHTS CHEMICAL WORKS
ERSKINE'S LOCATION OF HASENCLEVER'S "INTENDED WORKS"
JUDGE G.S. ALLISON
SAW MILL
FURNACE
Lake Tiorati
Little Cedar Pond
MT. FORGE LOT # 4
Pound Swamp
MINE
FINGERSTONE
MONROE
AND
HASENCLEVER MINE LOT
1000 A.
OLD COAL MINE
TOLL GATE
HAVERSTRAW
STORE
Rock House Hill
Beaver Swamp
DRAWN
C.K.T

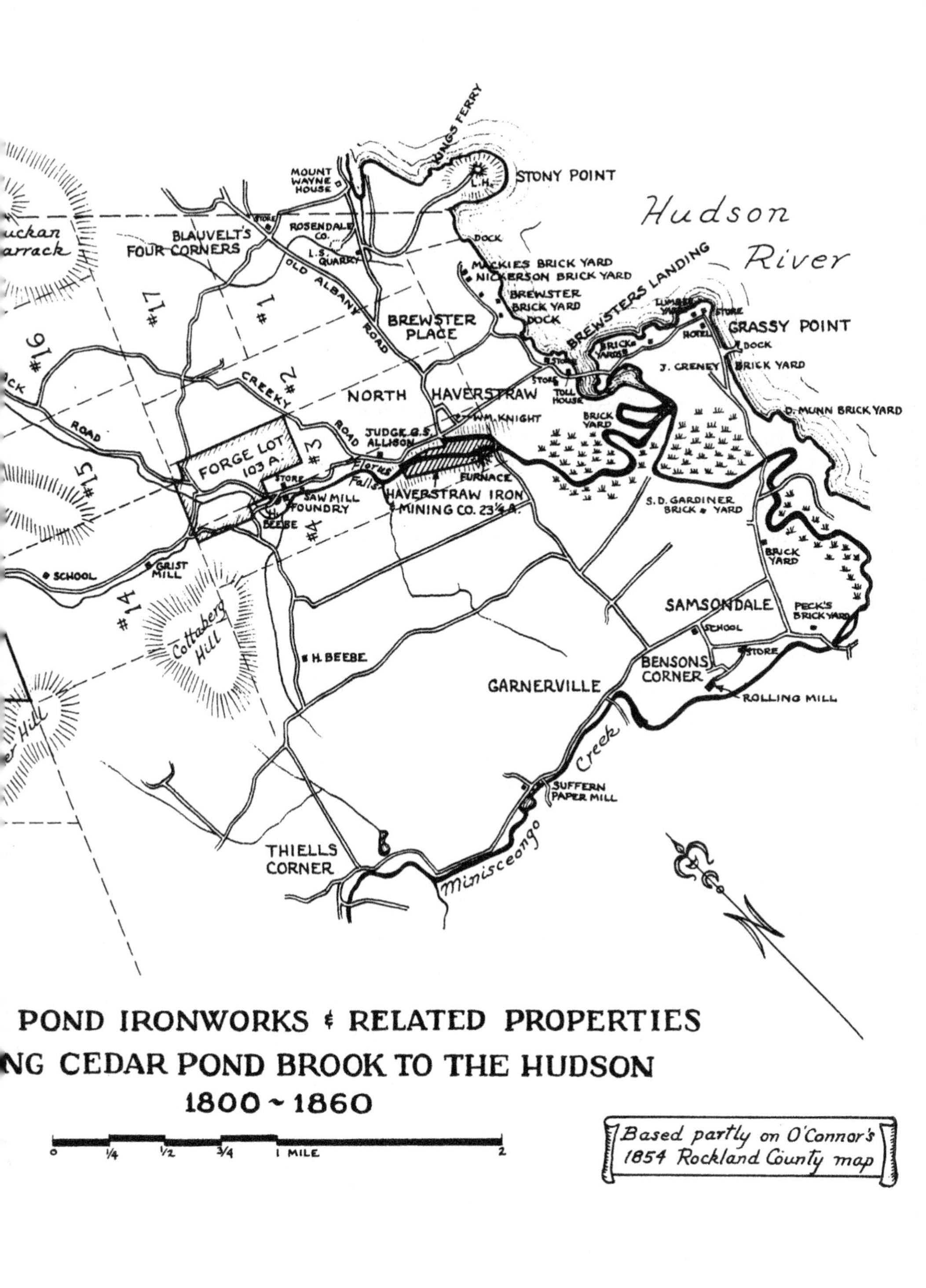
KINGS FERRY
MOUNT WAYNE HOUSE
STONY POINT
L.H.
Hudson River
STORE
ROSENDALE CO.
L.S. QUARRY
BLAUVELT'S FOUR CORNERS
DOCK
MACKIES BRICK YARD
NICKERSON BRICK YARD
BREWSTER BRICK YARD
DOCK
BREWSTERS LANDING
OLD ALBANY ROAD
#17
#1
#16
#2
BREWSTER PLACE
LUMBER YARD
STORE
GRASSY POINT
HOTEL
BRICK YARDS
DOCK
J. CRENEY
BRICK YARD
STORE
TOLL HOUSE
NORTH HAVERSTRAW
CREEK
ROAD
WM. KNIGHT
BRICK YARD
D. MUNN BRICK YARD
FORGE LOT 103 A.
#3
JUDGE G.S. ALLISON
Florus Falls
FURNACE
#15
STORE
SAW MILL
FOUNDRY
BEEBE
HAVERSTRAW IRON MINING CO. 23¼ A.
S.D. GARDINER BRICK YARD
#4
BRICK YARD
SCHOOL
GRIST MILL
#14
Cottaberg Hill
SAMSONDALE
PECK'S BRICK YARD
SCHOOL
H. BEEBE
STORE
BENSONS CORNER
GARNERVILLE
ROLLING MILL
Creek
SUFFERN PAPER MILL
Minisceongo
THIELLS CORNER
N
POND IRONWORKS & RELATED PROPERTIES
NG CEDAR POND BROOK TO THE HUDSON
1800 ~ 1860
0 ¼ ½ ¾ 1 MILE 2
Based partly on O'Connor's 1854 Rockland County map

mine during the Revolution, apparently with Livingston's approval. He also set up forges on Cedar Pond Brook.

In 1793 Brewster sold a forge and a portion of his property lying several miles down-stream from the Cedar Ponds to Christopher Ming. It is said that the old forge hammer could still be seen, not many years ago, lying in the bed of the stream. Ming sold the works and property in the same year to Halstead Coe, who three years later resold to John De La Montagne.

On November 17, 1799 Samuel Brewster sold Jonas Brewster half of a 664-acre lot, for $332. In later deeds this is referred to as the "Cedar Pond Furnace Lot" or "Furnace Tract" and included part of the Cedar Ponds and the stream flowing from them as well as Hasenclever's furnace site. Just about five months earlier (June 1) Jonas Brewster had purchased for $750 the 1000 acre Hasenclever mine lot from William Denning, Jr., and Thomas Hay of Haverstraw. It appears that Jonas and Samuel Brewster were preparing to work the Hasenclever mine and erect a furnace on or near Hasenclever's site on Cedar Pond Brook. The date of the building of the furnace has not yet come to light but the word "furnace" appears in later deeds for the 664-acre tract and gives the impression that 1800 would not be too far out of the way. The date the furnace was last worked is also somewhat nebulous, but Lesley in his *Iron Manufacturer's Guide to the Furnaces, Forges and Rolling Mills of the United States,* published in 1859, says the furnace was abandoned more than 40 years before. Lesley, as well as others, referred to it as the "Orange" furnace. The Greenwood, Queensboro, and Forest of Dean furnaces were sometimes called "Orange" also, and to add to the confusion the property changed hands many times after Samuel Brewster's death in 1821.[2]

In the summer of 1961, the ruins appeared to be in the same condition as they were a decade or two ago. A crude circle of stones, the remains of the crumbling walls, is all of the furnace itself which can be seen today. On the northern side, next to the former bellows arch, and still clearly defined is the waterwheel pit, 20 feet in length and slightly less than five feet in width. Parts of a salamander scattered around indicate probable trouble during the last blast.

Up the brook farther north is a picturesque cascade. This series of falls, a good 50 feet high overall, is fed by water from the dammed ponds and provided the power for the wheel to run the blast.

To find the site, follow the highway from Lake Tiorati's outlet down toward the Palisades Interstate Parkway and go on past the historical marker mentioning the ironworks. Almost half a mile from Tiorati's outlet, the ruins are located to the left, between the road and the brook.

DATER'S WORKS

Dater's works, established in 1800, consisted of two forges, one on each side of the Ramapo River at Pleasant Valley (or Pleasantdale), New York. These names have since disappeared and today the area is included in the northern section of the village of Sloatsburg.

Abraham Dater employed as many as 140 men in his ironworks, producing merchant iron (a bar form convenient for market) which was carted to Nyack, and then shipped by boat to New York City. One of his customers was Jeremiah Pierson, part owner of the Ramapo works, about two and a half miles to the south down the river. In 1806, Jeremiah wrote to Abraham, "We have concluded to 45 pounds, per ton, for iron of Sterling pig drawn to gauge, provided you engage to deliver us a given quantity, say 30 or 40 tons or the chief part of the iron you make." Dater used pig iron from Sterling and Southfield furnaces, which were operated at that time by the Townsends. The Southfield furnace had been put into blast just that year.

In 1812, Abraham Dater was the second-highest taxpayer in the township. The highest was Jeremiah Pierson, and John Suffern, third, all ironmasters. (The extent of Dater's holdings can be readily seen from the fact that he owned 2,600 acres between the Ramapo River and Stony Brook in Orange and Rockland Counties. On one lot of 35 acres he had a "good two fire forge, coal house, mansion house, store house, barn, cow house, smith's shop and seven tenements.") Dater made his son-in-law, Thomas Ward, a partner and the firm became known as Dater & Ward in 1820.

On July 8, 1823, he sold all but 750 acres of his property to Ward;

the remaining iron properties went to his son, Abraham, Jr., on August 7 of the same year. After the death of Abraham, Sr., in 1831, the works were sold to the Sterling Company, who ran them for a fairly short period. In 1849, they were taken over by N. P. Thomas, who was succeeded by A. H. Dorr, but Dorr held the works only a brief time before John Sarsen bought the forges, operating them until 1854, after which they were abandoned. Dater also owned the Split Rock forge on Stony Brook, about three quarters of a mile from Sloatsburg. In 1884, the split rock, from which the forge took its name, was covered by the construction of Allens Mill Dam.

The forge site located on the east bank of the Ramapo can still be found easily today, north of the bridge on the east side of State Route 17, two miles south of the Tuxedo Railroad station, where Washington Avenue turns east from Route 17 over the railroad. The east bank is littered with slag, charcoal, foundation stones, and the remains of an old dam. A natural rock ledge protruding from the shore at this point forms part of the dam, narrowing the river passage. The gentle slope is a favored place for deer to drink, as their tracks in the snow testify.

The name Dater has been perpetuated: a mile and a half to the northeast is Dater's Mountain with Dater's Mine on top of it.[1]

FOREST OF DEAN FURNACE

This furnace, in operation as early as 1756, was located about five miles southwest of West Point and a half mile east of Mine Lake, on Forest of Dean Creek in the Hudson Highlands.

In 1753, Vincent Mathews leased 4,290 acres "for the right to the Iron Oar standing, growing, lying, being found or to be found in or upon said tract." The following year, on July 30, he purchased this tract and an additional 500 acres from Nicholas Colton, who had bought it earlier the same day from Henry Case. On this newly acquired property, Mathews erected the Forest of Dean furnace.

The ore supplied for use in the furnace came from the neighboring Forest of Dean mine, one of the richest and most important in New York. In April, 1770, this furnace was offered for lease in the *New-York Gazette and the Weekly Mercury* by a John Griffiths whose impending return to England for business reasons was responsible. Among other information, the advertisement states that a source of ore is close to the furnace, water abundant even in the driest seasons, and wood for charcoal available within two miles. Griffiths also noted that since the distance from his ironworks to a water route was only five miles, iron could be transported to market for 40 shillings less per ton than from any other ironworks of which he knew. He rated the quality of the iron as above average, claiming it had been approved of and lately sold in Bristol, England, for 20 shillings more per ton than the finest Andover pig iron.

Samuel Patrick must have leased the works some time before the Revolution, for he was operating the furnace under this arrangement in the fall of 1776 and filling government orders for iron for the Continental Army. A letter written by General William Heath from Peekskill, New York, on December 2, 1776 to General James Clinton at Fort Montgomery places Patrick at the Forest of Dean furnace at that time. ". . . I have ordered some Sheet Iron to Fort Montgomery which is Designed for Stove Doors which are Casting by one Samuel Patrick at the Furnace of the Forest of Dean, you will please to acquaint him of it as soon as convenient." [1]

Patrick not only operated the furnace, but was also captain of a local company of militia in the Cornwall precinct known as the "Company of Militia of foot of Cants Hook," an assignment he received December 12, 1775.

His death probably occurred early in 1777 when the Convention of the State of New York appointed Christopher Tappen and William Denning to look into affairs at the Forest of Dean ironworks after Patrick's death. They reported back to the Convention on March 10, 1777:

> Your Committee appointed to confer with Mr. John Griffiths and Mr. Robert Boyd relating to the Hire of the Furnace of the Forrest of Dean, do report That Mr. Griffiths informs them that the lease by which Mr. Patrick held the Furnace Expires about the last of October next,[2] so that one blast only can be performed during the lease, but thinks it may still be continued in the Public service if found necessary, he also informs that the said Mr. Patrick died intestate.
>
> Your Committee have also carefully Exam'd the Estimate of the sundry articles &c., as per inventory and having taken the advise of Mr. Griffiths are of the opinion that the Prices annexed to the several

articles is their true value, Except the Water wheel, which your Committee thinks ought to be considered as part of the building. Mr. Griffiths further informs your Committee that he is of opinion the metal is good for Castings of all kinds, and that Two swivels were tried in N. York and proved good, of which a certificate was obtained. Mr. Boyd is of the same opinion, Mr. Griffiths is further of opinion that if the House should determine to try this Blast, it must be begun without delay, as the season for cutting wood and other necessary preparations will soon be past. Your Committee applied to Mr. Griffiths to superintend the said Blast, if the House should determine to prosecute it, but he declined it by reason of the unhealthy situation and his inexperience in Casting, and recommended Mr. Robert Boyd as a proper person for that business all which is humbly submitted.[3]

Three days later, Boyd and Griffiths petitioned the Convention of the State of New York with this offer, drawn up at Kingston.

May it please your Honors, As Forest of Dean Furnace is the only one in the State that can with convenience supply the Publick,[4] your Petitioners will put the same into Blast and use their Endeavours to make whatever castings they may be able to cast there on the same Terms that other Furnaces shall or will at the time of Delivery, provided they may obtain an exemption for Forty men for the space of seven months from military Duty, an order to be supplied with

a quarter Cask of Gunpowder   Two Hogsheads Melassus
one hundred Bbls of Flour   one Hogshead of New York Rum
four Hundred Bushels of Bran   Four Hundred Yards of Tow Cloth
Twenty Bbls of Pork   & one hundred pair of Men's Shoes, . . .[5]

It was George Leonard, appointed by the Convention in March, 1777, as administrator of Samuel Patrick's estate, rather than Boyd and Griffiths, who took over the ironworks. The connection which Boyd and Griffiths had with the Forest of Dean furnace evidently stemmed from the fact that the Patrick estate owed Griffiths money, according to this letter from Leonard to General George Clinton:

Orange Furnace March 22d 1777.

Hon'd Sir,

When Mr. Griffith[s] first Came up he Seemed in a Great fright Fearing he Should Loose his money, and Insisted to Know how he Could be

> paid. I told him we Could pay the Same Immediately, but it would take all my Running Cash, and prevent my being Able to Carry on, if in case the Convention should not take them; he Replyed if the Convention did not take them, he would Immediately, when they declined; I profered the works to him and Mr. Boyd, but Mr. Griffith[s] thinking Our Abilityes weak as to Cash, he then Made me an offer which seemed by computation about 1100 lbs. in his favour, this surprised me, I told him I would consider and see if I could better it; and finding the Convention very Honourable and willing to assist me I concluded to take the works in my Care, and sell some Pig iron to pay the debt and enable me to carry on, but he insisted of me to postpone selling the pigs, as he said he would wait for the money, and was sure pig iron would Rise, Likewise said he would leave the bonds in yours hands and give you orders to receive the same without the Intrest and Deliver the bonds when the principal was paid. I beg you would inform me if them are the orders, if not I must try to pay the same immediately as I dont choose to pay any intrest, as He has Always told me to depend he would never charge any.[6]

George Leonard seems to have run the works for at least a short time. From evidence given at a court martial held at Peekskill on April 13, 1777, Job Babcock stated that he had been working for him at the Forest of Dean furnace. The furnace most likely was in operation until October 6, 1777, the day that the British forces stormed and captured Forts Montgomery and Clinton. The proximity of this battle at the forts in the Hudson Highlands probably caused permanent abandonment of the furnace for no further references to production of iron here can be found. However, a 1781 map, drawn up by Thomas Milledge or Andrew Skinner for Oliver Delancey, shows the furnace site which is labeled on this map as "Colonel Mathews' Furnace." An earlier map of the area, made in 1779, also refers to it as "Mathews' Furnace."

The next mention to be found of the property is in an advertisement appearing in the *New York Spectator* of December 25, 1799. The owner at that time, E. Luget, expressing a desire to withdraw from business, offered it for sale. He described it as "The one undivided moiety of the forest of Dean tract, consisting in the whole 6600 acres of land." In the same advertisement Luget also offered the Queensboro tract and ironworks for sale, giving a detailed description of them. No further mention is made of the Forest of

Dean furnace, a tacit acknowledgment that it was no longer being worked and probably not even in operating condition. The property was not sold, for it was put up for auction on December 1, 1810, by order of the Court of Chancery, according to a notice in the *New York Evening Post* of September 14, 1810 as:

> All those two certain pieces of land, being parts of a certain large tract of land known by the name of the Forest of Dean Tract, and are distinguished in the subdivision of the said Forest of Dean Tract, made by William Thompson and Seth Marvin, Commissioners, under the Partition Acts by Lots No. 1 and 2, said lot No. 1 consisting of 1726 acres, and lot No. 2 containing 1654 acres.
>
> Also, all that certain piece of land known by the name of the old Furnace Mine, being also part of the said tract of land, called the Forest of Dean, and situated on and near Poplopes Kill, containing about 68 acres of land. . . .

Nothing further of interest can be found regarding this famous furnace site until March 27, 1942, when the property was bought from "The Forest of Dean Iron Ore Company" by the federal government to expand the land holdings of the U.S. Military Academy at West Point.[7] The land has been changed so radically that even the original contours have been altered in places. Many years ago an elderly local inhabitant, talking to the foreman at the nearby Forest of Dean mine, described the appearance of the old furnace site about 1790 as he remembered it. At that time, he said, there were "trees growing inside of the furnace." All that remains today are stories and legends like this to mark its memory.

GREENWOOD FURNACES

James Cunningham of Warwick, New York, highest bidder at the sheriff's sale, acquired on September 1, 1810 a tract of 3,400 acres comprising the northwest half of Great Mountain Lot No. 3 in Orange County.[1] He wanted the choice site for an ironworks, as well as for the abundant ore. The next year he erected the charcoal blast furnace, and named it for the rich color of the surrounding

Greenwood Ironworks. *The County Atlas of Orange, New York,* 1875.

forests. The furnace ruins can still be seen in a picturesque glen at the outlet of Echo Lake, formerly Furnace Pond, which lies just over half a mile east of the railroad station at Arden, New York.

According to contemporary accounts, a rousing celebration was held during the opening of the works with singing, many toasts, the roasting of an ox and workmen parading, each bearing the tools of his particular branch of the trade.[2] During the War of 1812, the works supplied cannonballs for the American forces and twelve workmen were exempt from military duty. The furnace was altered during the period since a cast iron plate over the tuyère arch bore the inscription "Rebuilt 1813." Business must have slowed down considerably, however, for the sheriff seized the ironworks on November 18, 1815.

There is a puzzling hiatus until 1827, when Gouverneur Kemble, owner of the West Point Foundry at Cold Spring, New York, and William Kemble of New York City acquired 2,839 acres of Green-

Greenwood Furnace January 2d

Mr Andrew Suffern Bot of John R Kimble & Co

Two Tons Pig iron one of which is to be taken Tomorrow $35 pr ton — $70.0

Note pinned in the Suffern Ironworks ledger, 1822. This is the earliest manuscript record from the Greenwood Furnace.

wood furnace property which had been "lately purchased by Nathaniel Sands at a Sheriff's sale." [3] Gouverneur had been leasing land in Orange County for the mineral rights as early as November 11, 1823 in order to insure a good supply of ore for his foundry. Before much longer, the two Kembles and a silent partner, John R. Fenwick, owned most of the mines in the Greenwood region, as well as the Greenwood ironworks. Fenwick sold his one-third interest in the "Greenwood Furnace and all machinery, buildings, stock" and land to Robert P. Parrott for $10,000 on February 16, 1837.[4] Parrott, an 1824 West Point graduate, was commissioned captain in the ordnance corps in 1836 and assigned to active duty at the foundry just across the Hudson from the Military Academy. Shortly after, he resigned from the army to become superintendent of the foundry for the Kembles.

In 1837, his brother Peter, a former whaling captain, came to Orange County to manage the Woodbury furnace for Gouverneur Kemble, and before long he was also managing the Greenwood furnace. He had some bad luck in February of 1839 when all of the woodwork about the furnace, except the waterwheel, was destroyed by fire but in rebuilding the charging bridge and the bridgehouse, the furnace was increased from its former height of about 36 feet to 42 feet. New blowing machinery was installed at the same time. The improvements raised the production of the furnace to about 30 tons of iron per week. On December 1, 1839, Gouverneur and William Kemble sold their combined interest in the Greenwood ironworks, "it being ⅔ of the premises," to Robert Parrott for $20,000. The two Parrott brothers were then in full control of the ironworks and worked successfully together for many years.

Construction of the Erie Railroad through the Ramapo Valley to

Greenwood in 1843 was a great advantage; another was the discovery of promising new veins of ore. Business improved to such a degree over the next ten years, that a second furnace, called the "Clove" or "Greenwood No. 2," was built. Unlike the first Greenwood furnace, it used anthracite for fuel rather than charcoal. This water-driven unit, 54 feet high, 18 feet wide across the bosh, 37 feet at the square base and open at the top, went into operation on July 1, 1854.[5]

This furnace principally made foundry pig iron, stamped with the brand name "Clove," for fine hardware and stoves. In 1855 it produced 5,000 tons of iron. Most of the ore used in it came from the Greenwood group of mines and was roasted in kilns to remove sulphurous content before being put into the furnace. The kilns were located about a half mile up the brook from the old charcoal furnace. Here also was a stamping mill where the ore was broken into small pieces. Lime was not used in the smelting at the furnace as the ore was self-fluxing.

Prior to the Civil War, the charcoal pig iron from Greenwood furnace No. 1 had been selected by the West Point foundry for the manufacture of gun barrels because of its remarkable strength as displayed in government tests. Robert Parrott had kept in close touch with ordnance requirements of the federal government, designing and perfecting a hooped gun which was ready for use when the Civil War broke out.

This remarkable piece of artillery became famous during the war as "the Parrott gun" and was credited with being the most effective artillery weapon of the Union Army. In fact *Frank Leslie's Illustrated,* May 10, 1862, a well-known weekly, devoted the entire upper half of their front page to an illustration of the Parrott gun, giving both a side and rear view of it.

Parrott, himself, supervised the manufacture of the guns at Cold Spring and because of the outstanding record that these guns made in battle the Greenwood furnace was kept busy throughout the Civil War.

Robert's correspondence from the foundry at Cold Spring to his brother at Greenwood indicates his great concern for the important work that was being carried on at both places. "My interest in everything at Greenwood is very great, but under all the circumstances in view of the great extent and importance of the operations here, I find it difficult to leave. We continue very busy and

FRANK LESLIE'S ILLUSTRATED NEWSPAPER

Entered according to the Act of Congress in the year 1862, by Frank Leslie, in the Clerk's Office of the District Court for the Southern District of New York.

341—Vol. XIV.] NEW YORK, MAY 10, 1862. [Price 6 Cents

The Parrott guns made the headlines.

likely to use all the iron you can make." Two days later, on April 25, 1863, Robert emphasized the demand for Parrott guns saying, "Guns are ordered by the fifties and all my efforts are required to keep the supply." And on May 2, "I am very anxious about the great increase in demand for our work and it requires no little thought to provide for it. Unless some material change comes, I must look for more extensive preparations of some kind, otherwise matters are getting on well. I hope to be over very soon." [6]

Ovens were used at Greenwood No. 1 to heat the blast. This method required less charcoal in the furnace to smelt the ore, as compared to the cold blast. Some customers, including the government, insisted on the cold blast because of a belief that the best iron was produced in winter. To make certain that this cold blast method was used for the guns ordered, naval Lieutenant Walker was stationed there. The foreman in charge related in later years that he used to run the furnace on hot blast by keeping the valve to the heating ovens open until he saw the lieutenant coming around the turn in the road to the cinder dump. He would then close the valve and run it on cold blast. The officer never discovered the trick, and the quality of the ore never suffered, according to the foreman.

An interesting description of the furnace during the busy days of the Civil War period is given by Richard D. A. Parrott in his memoir published in 1921:

> There is little remaining now to indicate the activities of that earlier period. The population dependent on the business was about fifteen hundred. Houses of the furnacemen, miners, and woodchoppers were scattered through the mountains. At the furnace pond the houses were numerous enough to make quite a village. Traveling eastward from the Ramapo Valley up the public road along the ravine in which the brook flows and after passing the Greenwood Furnace, which looked formidable because of the enormous charcoal sheds crowded near it, one came upon more sheds on the right, over near the pond; straight ahead on the right side of the road was the Mc Kelvey house and further along on the opposite side was the store, housing the post office and carrying everything from calico to pork; then came several dwellings on both sides of the road and finally the school house on a hill at the left.
>
> The burden of the furnace was made up of ores mined on the property; those mines contributing at different times to the gun metal were the Bradley, Greenwood, Surebridge, Pine Swamp, O'Neil and Clove. The ores were not hauled directly to the furnace, but to a kiln a half a mile above it, in which they were roasted to drive off the sulphur. After this they were put through a stamping-mill which reduced the size of the lumps to that of a pigeon's egg. The brook from the upper ponds was dammed at this point, giving power to run the stamping-mill and also a sawmill.[7] The wheel used for this power was of Scotch design, the first turbine wheel to be put in place in this country.
>
> An important branch of the business was the cutting annually of ten thousand cords of wood and turning it into charcoal. The hauling

of this coal from the pits and storing it in sheds for use during the winter, required a great deal of labor. Numerous baskets were needed and these were made at their homes by the mountaineers or so called "bocky-makers." There were no narrow gauge railways or wire rope tramways to handle materials for the furnace. Transportation of these meant the maintenance of numerous teams of horses and mules and required that constant attention be given to the roads. In spite of this the roads were noticeably poor and not conducive to pleasure driving.

The community of course had its military company. The men were in uniforms of blue flannel and their guns were of wood. By drilling and marching they got a notoriety that went over the top of the mountains to West Point. One evening Captain Alfred Mordecai appeared on a recruiting tour and addressed the raw soldiers from the piazza of the Greenwoods, exhorting them to enlist in the service of their country. His speech was loudly cheered and he continued: "I can tell by your voices where your hearts lie." This appreciation aroused "pep" in the men and one-third of the company heeded his call and went off to war. So, it was not long before Greenwood had tears and shed them at the cemetery and it also had a returned soldier carrying an empty sleeve.

The free use of the land for grazing cattle; the opportunity for securing free wood for domestic use as well as for making baskets, spoons, ladles and trays; the unrestricted use of the woods and ponds for shooting and fishing, all tended to make the mountaineer indifferent to discipline. His labor was irregular owing to his independence. The village was "dry" by regulation, but nearby centers were "wet" without regulation. The consequence was that walking, to and fro, was a compelling pastime and personal liberty was in full swing. The new freedom of the present day pressed but lightly then, if at all.

A visitor would have pointed out features of the place that the inhabitants overlooked as being commonplace. He would have said that much of the charcoal escaped being fed into the furnace for it was being wafted about in clouds of dust settling on everything that couldn't get out of the way—including the face of the visitor. By nature the local color was green, but artificially, black was a good second and more general worn. The visitor would have asked "Why this noise, in an otherwise perfectly good village?" He would have been told that it was the stamping-mill and, though loud, was thought to be harmless by the natives. Speaking of noise, when the guns increased in calibre to the 200 pounders, the firing tests at Cold Spring, twenty miles away, were distinctly heard, reverberating over the Hudson Highlands, in this village. The casual mention of this foundry incident does not convey the satisfaction that was felt by those connected with the furnace. It confirmed in their minds the thought that they were personally engaged in a patriotic work.

[These were extremely busy and anxious days for the Parrotts.] The occupation was not an indoor one for the owners. Mines and coaling jobs had to be visited by day and the furnace at night. Midnight casting was something worth staying up for. As often as circumstances would permit, Robert would come across the mountains from Fort Montgomery, calling at a mine on the way, in hope of finding Peter there. Although they often had interviews, they also exchanged letters with great frequency. The undertaking before them, was the putting down of the Rebellion and they followed the first years of indifferent success of the armies of the North with great anxiety.[8]

When Greenwood furnace No. 1 completed its final blast in September of 1871, the end came for the charcoal blast furnaces in this area. Situated in a ravine, the ruins of this furnace are well worth a visit. The arch on the western side is off-center and leans to the left, a rather unusual construction. The stack, as might be expected, has deteriorated over the years and the highest wall now standing—about 25 feet—is beginning to crumble. Two fair-sized hemlocks have grown in the narrow space between the stones above the arch, and give the impression of being rooted in the rock itself. The 25-foot square base is now covered with fallen stones and debris from the upper part.

Greenwood furnace No. 2 continued in service after the closing of the old No. 1 charcoal furnace. It was not shut down until 1885, when the fire was allowed to die out, even though the furnace was still filled with ore and coal. One continuous blast had run from May 26, 1871 to June 14, 1881, making 101,245 gross tons of pig iron, 75 percent of No. 1 grade.[9] Anthracite was the fuel used, but during the last three years this was sometimes alternated with a small quantity of coke.

Mineral wool—an incombustible nonconductor of heat—was something new in the iron world. England was making it and calling it "silicate cotton." The Krupp works in Germany called it "slag wool." The first commercial production in this country was manufactured in 1875 at the Clove furnace at Greenwood.

The process involved subjecting a small stream of molten slag to the action of a jet of steam. The effect was that of spun fibers

Ruins of the Greenwood Charcoal Furnace. *Photo by the author, 1940.*

of wool or cotton. Greenwood produced this for seven to eight years.

Peter Parrott had already bought up his brother's share of stock some time before Robert died in 1877. The total holdings were approximately 10,000 acres—33 separate acquisitions altogether. Back in 1864, the total was only 8,505 acres, but this had been increased by 20 purchases, among them the Forshee, Mombasha, Bull, Clove, and Warwick mines.

After Robert's death, the company still continued to use his name on its letterheads, but like its counterparts, the Parrott Iron Company gradually drifted into idleness and, at last, into oblivion. A year after the closing of the Clove furnace, the Greenwood property was in receivership to Stuyvesant Fish of New York, whose agent, C. M. Nichols, had worked at one time for the Parrotts.

Ruins of the Clove Anthracite Furnace at Greenwood. *Photo by James Skyrm.*

Offered for sale, it failed to find a buyer and was finally partially dismantled.

The ruins of the Clove furnace, about 200 yards east of the railroad station at Arden, New York, are in a fair state of preservation. The furnace with three tuyère arches, the casting arch, and also part of the trestle from the summit of the hill to the top of the furnace are interesting examples of how old ironworks were built.

Nature has taken over: the soot from the ironworks, which once covered every inch of ground, has given way to the magnificent beauty of the glen and its surrounding hills and the name Greenwood has resumed its original significance.

MONROE WORKS

In 1808, when the town of Southfield, New York, changed its name to Monroe,[1] Joseph Blackwell and Henry W. Mc Farlan(e), who were then erecting an ironworks in the Southfield area, decided to adopt the new town name for their enterprise. These works, consisting of a rolling mill, a nail manufactory, and two trip-hammers, were located on Mombasha Creek adjacent to present State Route 17. Here they produced hoop iron and cut nails from threepenny to twentypenny in size, all warranted equal to any made in this country.[2]

Blackwell and Mc Farlan, partners since 1796, were iron merchants in New York City, where they did an immense business from their store at 8 Coenties-slip. They were known to buy out entire cargoes of iron without hesitating, and their store was considered the headquarters of the iron trade. They advertised in the *New-York Herald* of March 31, 1804, that they had for sale pig iron from Ringwood, Martha, Weymouth, Livingston, and Salisbury furnaces; casks of fine rosehead English nails, as well as American nails and spikes; a very large assortment of Swedish and Russian refined and bloomed iron; anchors weighing from 100 to 500 pounds; plus forge hammers, anvils, kettles, shovels, scythes, and other implements.[3] A year later they were selling cut nails and wrought nails made at the state prison.

With the building of their Monroe works, the owners were no longer dependent on others for their supply of nails and hoop iron. Before many more years passed, they also acquired a forge and furnace at Dover, New Jersey, formerly belonging to Israel Canfield and Jacob Losey, who failed in the depression following the War of 1812. Few facts can be found about the number of employees, working agreements, or the actual physical appearance of the Monroe works during its early years. Blackwell died in 1827 and Mc Farlan in 1830, with Henry Mc Farlan, Jr., then taking over. Production figures for the year 1832, taken from Thomas Gordon's 1836 *Gazetter of the State of New York,* show that it was a fairly large operation employing 100 workmen, using 20 tons of anthracite, 30 tons of bituminous coal, and 1,500 cords of wood to produce 420 tons of nails and 500 tons of hoops.

On April 21, 1837, Lewis C. Beck, a professor of chemistry and natural history at Rutgers College, visited the Monroe works and entered in his diary:

> Monroe Works three-fourths of a mile from Southfield Furnace, under the direction of Hudson Mc Farlane, formerly of Schenectady. Here was a rolling mill, shovel factory etc. which was burned a year or two since. With him we visited a locality about three miles distant called the Silver Mine [owned by Mc Farlan], an old digging supposed to have been made during the Revolution. From this locality we passed over the Black Rock Mountain by a circuitous route of four or five miles to the Rich Iron Mine [also owned by Mc Farlan]. This mountain is the spine of the Highlands in this part of the Country. The Rich Mine was formerly worked to some extent. Its ore is highly magnetic. On the summit is one large stone, seven feet high and 27 feet in diameter, resting on five detached fragments of rock. There can be no doubt that natural agencies have brought it to its present position and yet it has every appearance of being the work of art. Returned to Monroe Works and lodged at Major Wilks. Beautiful aurora this evening with a rose tint.[4]

Samuel Eager wrote in 1846 of the Monroe works: "The establishment is in full operation and is vigorously prosecuted by its present owner, Hudson Mc Farlan, Esq." The Monroe works were still active in 1882, but seem to have been abandoned shortly thereafter.

A rare old photo of the Noble Furnace, circa 1870, before the south side and arch caved in.

NOBLE FURNACE

William Noble of Bellvale, New York, acquired by patent from the State of New York on April 29, 1813, a strip of 870 acres along the west shore of Greenwood Lake. Here, about 150 yards inland, on a stream now called Furnace Brook, he built a furnace about 1833.

The furnace was a failure from the beginning, because the stream —only a few feet wide—could not supply sufficient waterpower for the blast. In 1939, John Townsend, with a remarkable memory at eighty-two, and a good knowledge of the area, told the author that the furnace was called the "Succor" or "Sucker" because everyone who invested in it lost his money.

One can still see the ruins of the furnace, its walls composed mainly of thin shale rock.[1] Consideration of the location on the small brook makes one wonder why such a site was chosen. The answer seems to be that the construction of a sizeable reservoir had been planned above the furnace. But the contour of the land on the side of the hill, and the size of the stream hardly warranted the effort, assuming the brook was as small in the early nineteenth century as it is today.

QUEENSBORO FURNACE

The restored ruins of this old furnace are still standing at the junction of the Queensboro and Popolopen Creeks, about two-and-a-half miles west-southwest of Fort Montgomery, New York.

The date of its erection is unknown. Most historians feel that Queensboro was active during the Revolution, but it seems unlikely that it was built before 1783. Two things have led to the confusion and mystery which surrounds the date. First, it was sometimes referred to as the Orange furnace, rather than by its correct name of the Queensboro; also the Forest of Dean furnace, just to the

The east side of the Noble Furnace, showing the deterioration of the south side. *Photo by Vernon Royle.* (See previous picture.)

north, was at times also called the Orange. This double identity has become a stumbling block for many researchers, since military correspondence in the early years of the war sometimes referred to the Orange furnace. However, this reference was to the Forest of Dean, in operation at that time, rather than the Queensboro.

The second reason responsible for the belief that the Queensboro furnace was standing in 1777, is a plaque on the north side of the furnace which reads:

ORANGE OR QUEENSBORO FURNACE
Erected 17–
On the 6th of October 1777
A British Column of Nine Hundred Men
On their way from Stony Point
To Attack Fort Montgomery
Forded Popolopen Creek
At This Furnace

This tablet is misleading. No furnace is shown at the Queensboro site on any of the maps of the period, whereas the Forest of Dean furnace does appear on some of them, as well as in military correspondence.

Ownership of the property can be traced back to October 18, 1731, when the tract of 1,437 acres where the furnace now stands was granted to Gabriel and William Ludlow. On February 13, 1775 the following advertisement appeared in the *New-York Gazette and the Weekly Mercury:*

> To be sold, on the premises, the 25th of March next, at public vendue, if not before disposed of at private sale, A VERY VALUABLE TRACT OF LAND, situate in Orange County, about 50 miles from the City of New-York, and two miles and a half from Hudson's river on the westerly side thereof, containing about two thousand acres and known by the name of Queensberry . . .

This early advertisement also states that the tract was "uncommonly well watered" and "heavily timbered"; the only buildings on the property were a sawmill and three houses, one of them new; there were two landing places that could be used by the purchaser, one

Queensboro Furnace. *Photo by Vernon Royle, 1921.*

being at "Poplope's Kill," which was also the landing place for the "Furnace of Dean," and the other on the west side of Salisbury Island where there is "a good waggon or cart road to Queensberry."

No mention is made in this advertisement of any ironworks here at that time, but the Forest of Dean furnace receives prominent mention in the latter part of the notice in this manner:

> . . . The above premises are in good repair and happily situated in a thick settled country, having the Furnace of Dean within two miles and a half of the principal dwelling-house, which will always provide a ready market for a great part of the produce of the farm, besides the convenience of a weekly conveyance to New-York during the season. For further particulars enquire of Mr. Robert Ross at the North-River in New-York, or of Moses Clement, Esq., on the premises.

The earliest mention of ironworks, which can be in any way construed as being on the Queensboro tract, is by Hector St. John de Crèvecoeur, who visited the area about 1789. Writing about Popolopen Creek, he says: "It is only after having put in motion the hammers of two big forges and the bellows of two furnaces, known under the same name, that this little river joins its waters with those of the Hudson. . . ."[1] Unfortunately Crèvecoeur has proven to be somewhat unreliable, drawing upon his imagination now and then.

A gap occurs in Queensboro's recorded history from 1789 until Christmas Day 1799, when an advertisement appeared in the *New York Spectator*. Here is the first contemporary description of the Queensboro ironworks to be found in print that gives an overall picture of the properties:

> QUEENSBORO' IRON WORKS FOR SALE
> . . . consisting as follows:
>
> The Queensboro Tract, containing 1,437 acres of good meadow, pasture and woodland. The one undivided moiety of the forest of Dean Tract, consisting in the whole of 6,600 acres of land as above.
>
> The Bearhill Tract, containing 800 acres, mostly wood-land, and some meadow ground: And
>
> One undivided share of another tract of land called the Staat's patent, containing 400 acres of tillable and woodland.

After giving the location in relation to New York, the advertisement points out that these properties can be purchased either as a unit or in two separate parcels. The Queensboro and Forest of Dean tracts are joined to form one parcel and the Bearhill tract and Staat's patent combined to form the other group. Interestingly, the Queensboro and Forest of Dean were under the same ownership. Also at this time, iron was being made at Queensboro furnace but not at the Forest of Dean which was in ruins. The Forest of Dean mine was still being worked and supplied the Queensboro furnace with ore.

> The two first tracts, the Queensboro' and the forest of Dean, on which are erected the Iron-works, consisting of a furnace and forge, with two fires, both in compleat repairs, and provided with all the buildings necessary for carrying on the said Manufactory, viz. coal houses, carpenter and blacksmith shops, a convenient two story house for a manager, built last summer, and a number of other frame and log houses, with gardens, for the accommodation of the workmen—there is besides, a saw mill erected near the furnace, which supplies the establishment, and the market, with a considerable quantity of lumber; and a farm improved on the Queensboro' tract, equal for grazing to any in this state, with all necessary out houses . . . the quality of the metal is perfect—the agents for government, the states of New-York and Connecticut have drawn considerable supplies of ammunition from the Queensboro' furnace to their avowed satisfaction—the short distance of 2½ miles from the works to the landing is a considerable advantage, and the average benefits of the said works can be demonstrated to any person desirous of purchasing.

Further description surprisingly reveals that one of the earliest nail manufactories in the nation was located at the Bearhill tract, a discovery previously overlooked by historians:

> The other two tracts, the Bear-Hill and Staat's patent, tho' bounded on the two former ones, may be considered as a distinct establishment and purchased separately, they run 2½ miles along side of the river and comprehend a commodious creek [Popolopen], navigable for vessels of 50 tons—here are erected on the said creek, three docks, three dwelling houses and a store, a grist mill, a saw mill and a slitting mill, wherein a nail manufactory has lately been established . . .

Following a description of an elegant two-story house just erected at the spot where Fort Clinton formerly stood, on the south side of Popolopen Creek, and the plentiful fish pond (Hessian Lake) near, the buyers are promised indisputable titles to the properties and are directed for further particulars to either E. Luget, on the premises, or C. Lagarenne, 6 Duane Street, New York.

When it was decided to put the Queensboro furnace out of blast in 1800, some of the workmen, learning of this decision, left before the actual shutdown to seek new jobs.

Ten years later, on December 1, 1810, the property was again put up for sale at public auction. This time the notice appeared in the *New York Evening Post* of September 14, 1810. The Queensboro tract was still, as in the 1799 sale, just one of four properties and is described as:

> all that certain tract of land known by the name of Queensborough, lying being near the Dunderberg Hill, and situate on and near a certain Brook called by the Indian, Sickoftenskall, and by the Christians, Stony Brook, which Brook runs into the Poplopes kill, containing 1437 acres of land. . . . [A description which followed indicates that changes had taken place in the ten year interval] On the premises are . . . two large forges for making bar iron, each forge having two fires—one of them is not yet quite finished, but the one that is completed, has made, with one fire only, two tons of iron in a week. One of the forges is about 300 yards from an inexhaustible iron ore mine, the other about two and a half miles from it, approaching the North River; the mine is about five miles from the river, its ore is of excellent quality, very abundant, and so rich as to furnish a ton of bar iron from two and a half tons of ore.

The mine "about five miles from the river" no doubt was the Forest of Dean mine. Note is made that more ore can be mined than will be needed to supply the two forges, estimating that 2,000 tons could be sold annually.

Mention is also made of a mine within the "Queensborough" tract, whose ore, when mixed with Forest of Dean ore, makes excellent castings. Only one reference is made to the Queensboro furnace, which had been out of blast for ten years. "There is the stock of a furnace at a short distance from the lower forge, which could at a small expense, be put in good order." Nor were the mines, furnaces, and forges the only useful and profitable attractions at this Popolopen Creek site.

> A new grist mill on Polopes Kill, near the North River, dimensions 65 by 45 feet, having four pair of best burr stones of five feet and is believed to be capable of manufacturing 20,000 barrels of flour in a year. Boats of upwards of 50 tons may load and unload along side of the mill. There is a two story dwelling house near the mill, with a small dwelling house and a cooper's dwelling house and a cooper's shop, 60 feet by 20, conveniently situated.
>
> Also, on the same stream, an old Grist Mill with two pair of stones; the frame and boarding of this mill are sound. . . . Near the mills, on the same stream, is a new saw mill with two saws, capable of cutting into boards 1000 logs the season—the mill is provided with machinery to draw the logs into it from the water—farther distant in the woods is another saw mill, with one saw, situated among fine timber.

The sale of these properties held "in pursuance of an order of the Court of Chancery" (Samuel Corp, John F. Ellis and Gabriel Shaw, versus Alexander Macomb and others) was signed by Thomas Cooper, Master in Chancery, May 16, 1810.

Professor Beck visited the area in August of 1838 and noted that George Ferris owned a forge at Queensboro, as well as the Forest of Dean mine at that time.

The forges on this property were worked for a number of years, one of them still being active as late as 1843. The furnace, however, was never used again. In 1885, a writer named Kirk Munroe, visiting the Queensboro furnace ruins, described them as one of the most picturesque to be seen in the country:

> It stands in an open field by the roadside and is about thirty feet high, its grey walls are mantled with ivy and from its crumbling crest springs a clump of good sized trees. Its arched entrance and interior dome are clean-cut and unbroken as when the builders left them and are beautiful specimens of the stone-mason's art. It is known as the old furnace "par excellence," and is said to have been erected under the personal supervision of Kosciusko, the gallant Pole.[2]

The furnace was restored at federal government expense in 1912 and is once again in good condition. The most unusual and significant feature is its unusually high, sharp-pointed archway on the south side. The property on which it stands was acquired by the United States Military Academy in 1942 and is not open to the public. W. A. Rigby, who had been superintendent at the Forest of Dean mine, believed the old furnace capable of producing about one ton a day and he thought that it probably required a number of men to operate it. From the many pieces of slag found around the furnace, he surmised that it must have been difficult to keep the furnace operating smoothly. Remains of old charcoal beds

Queensboro Furnace. *Outing and the Wheelman,* Dec. 1884.

can be found in the surrounding woods, and the hillside back of the furnace is covered with charcoal debris, according to Rigby.[3]

Unique Popolopen Creek with its natural beauty and the historic interest of the ironworks is well worth visiting today. Thousands of mammoth boulders from the Glacial Age lie around the area as if tossed by a giant attempting to block the flow of the stream to the Hudson. Walking in this area, particularly along the north bank of the creek from the furnace site to the river is rough but rewarding because of the magnificent views of the rugged surroundings.

Nearly three-quarters of a mile downstream from the old stack on the north bank of the creek is the site of one of the forges. In 1958, a broken iron pig was found almost totally buried at the water's edge, with the letters "SBORO" on it. The portion marked QUEEN was missing, however. This fragment is important, as proof that the name was spelled "Queensboro" rather than "Queensborough" as it often appears on today's maps. The old foundations of the forge can be seen still. Years ago according to King Weyant, former caretaker for the Queensboro property, who remembers those days vividly, a scrap dealer collected the loose iron, but fortunately he missed this shattered piece which is now a treasured possession of the author.

There are many other traces of Queensboro's history still visible: the outline of an old road that dipped down near the creek and then crossed it; the ruins of an old mill with its raceway running through an overhead flume along a hillside to the mill wheel as shown in a nineteenth century Currier & Ives print; and the scattered rubble marking positions of other buildings which once were used to house men and equipment. All is now reverting to that natural state in which those pioneer ironmasters must have found it when they started to create the once thriving hamlet of Queensboro.

RAMAPO WORKS

In the spring of 1795, Josiah G. Pierson, owner of a nail factory in New York City, came to the Ramapo Mountains.[1] About a year before, he had received a patent for cut nail machinery and he was looking for a favorable site on the Ramapo River where he could build a rolling and slitting mill, as well as a nail manufactory. As the nearest rolling mill was in Wilmington, Delaware, he decided to build one nearer where the iron could be rolled into sheets of the proper shape. Impressed by the advantages offered he promptly purchased 119 acres of land from John Suffern in what is today called Ramapo.

Construction got underway in the latter part of May, 1795, after the first tools and supplies were sent from New York. That October, Josiah and his brother Jeremiah gave the new place the unimaginative name of "Ramapo Works." They built a 120-foot dam followed by sluiceways, a blacksmith shop, and a rolling and slitting mill. A group of buildings for cutting and heading nails all under one roof, measuring approximately 100 feet by 150 feet, was also put up.

Jeremiah Pierson moved to Ramapo June 1, 1796 to supervise construction while his brother remained in the city to handle business matters including the shipping of necessary supplies to the new works. Manufacturing operations were started in the spring of 1797. When Josiah died in December, his brother Isaac took over management of the firm's New York interests. The original name of the firm, Josiah G. Pierson & Bros., was retained until 1807.

In the spring of 1798 the 65 hammer and nail cutting machines used in the New York factory, were shipped to the new works together with 31 heading machines. By July all manufacturing operations, including nail cutting, were under way at Ramapo.

The demand for nails and hoops used to make sugar barrels for the West Indies trade was so great that the Piersons found a brisk business right from the start. A community store was also built in 1803. It became a shipping station after the coming of the railroad. By 1806 the demand for nails and hoops was still so great that 142 men were employed and the Ramapo works had added a forge, sawmill, and gristmill to the buildings previously constructed. A post office called "Ramapo Works" was established the following year which supplanted an earlier one just a mile or so south in New Antrim (Suffern). In 1809 Jeremiah Pierson, writing to a correspondent, described the various types and amounts of iron used at Ramapo and the proportion which was native ore. "Of the iron used here three-fourths is Russian and one-quarter is made in the vicinity. There are sixteen forges in this County [Rockland] making 300 tons monthly and in Orange County 600 tons are made, three-fourths of which is refined iron."

By "vicinity" he referred to neighboring ironworks at Augusta, Pompton, Southfield, Sterling, Ringwood, Suffern, and Pleasant Valley. In 1810 the Piersons employed only 150 persons at their works, but they were the means of direct support for at least 400 people. In addition, those employed part time such as woodcutters, charcoal burners, and natives of the surrounding Ramapo Mountains who sold their carved wooden ladles, hoop staves, baskets, and other hand-crafted articles to the Piersons, also received money from the Ramapo works. During this year, no less than 700 tons of iron were rolled and slit, with 500 tons of these processed into nails. Keeping pace with this production growth, the little community surrounding the works increased its number of buildings to 60, including a newly-formed Presbyterian church.

The War of 1812 brought demand for iron products of many kinds and continuing prosperity to the Ramapo works during the early nineteenth century. The Piersons also expanded their product lines by producing, in addition to nails, woodworking files. Jeremiah Pierson in 1814 gave this brief but comprehensive account of supply difficulties and general business conditions in his correspondence.

> We manufacture annually 500 tons of nails. Since the War prices have advanced caused by the difficulties of importing bar iron, for it is an undeniable fact that the quantity of bar iron made in the United States does not exceed one quarter of the consumption. The nails we make comprise all sizes from one inch to six inches long, together with floor brads and spriggs. Foreign steel is also an article of prime importance in the manufacture of our nails.

Two years after the peace treaty had been signed at Ghent, the Piersons branched out into a new business venture. A cotton mill five stories high measuring 140 feet by 40 feet was completed at Ramapo and went into operation. Spun yarn, a commodity in demand in Russia, could be exchanged there for Russian iron which had the high quality needed. Apparently the new mill failed to live up to advance expectations at first, according to a letter written at the Ramapo works on May 28, 1817, by John Gibson, otherwise unidentified:

> I take this opportunity to inform you that I am in good health at present hoping these few lines will find you in the same condition. I am at work at Mr. Piersons at present but I do not expect to stay long as the cotton factory is nearly stopt and business is getting dull when I get discharged I calculate to go to the westward on toward the Genissee country.
>
> When you get this letter I want you to go to Millers immediately and let them know and rite a letter and send it up so that I can have it on Saturday evening for perhaps I shall go away on Sunday and then you wont now where to send it. Direct it John Gibson ramapow piersons factory. I want you to let me know a little about the last scrape I had in New York no more at present.

This letter was addressed to Henry I. Traphagen, Hoboken, Bergen County, New Jersey, but nothing else is known about this communication.

By 1820, however, the cotton factory was running smoothly, employing 119 people. In the same year 700 tons of iron were rolled and slit and 500 tons made into nails. Jeremiah Pierson filed papers two years later for the incorporation of the firm under the name of The Ramapo Manufacturing Company. The same year, Isaac sold his half of the stock in the company (200 shares) for $145,000 to his brother Jeremiah, who now became the sole proprietor. The firm employed about 300 men at this time, and gave direct support to about 700 people.

Considerable attention was devoted in 1830 to the German or cementing process for making blister steel used in wagon springs. Swedish bar iron was used for this purpose. Covered by a domelike chimney, the cementing or converting furnace consisted of two narrow chests or boxes made of firebrick, each 15 to 16 feet long, 3 feet wide and 3 feet deep with small flues through which the flames could pass both between and outside. The firebricked boxes were filled with alternate layers of ground charcoal and bars of iron and the tops cemented over to exclude the air. Fires were lighted beneath them which were kept burning for 10 to 12 days. The blistered steel was then taken to the mill and rolled into shapes known in the markets of that day as "Pierson's spring-steel." Because of booming business another steel furnace and factory for the manufacture of screws was built in 1832.

Although no real basis for comparison between the number of

employees at the Ramapo works during 1832 and 1835 exists, there were 100 men working in the mill and cut nail factory and 200 in the cotton works during the later year. The village population had reached a figure of over 800 by this time, most of whom were housed in neatly painted, wooden homes on the east side of the Ramapo River beneath the sharply jutting ridges of the Ramapo Mountains.

When Lewis C. Beck visited the Ramapo works on September 7, 1836, he witnessed the cutting of a three-cornered file which he described in his diary. The iron was first softened, then cut with a chisel and afterwards hardened, the whole operation being performed by hand. Professor Beck was impressed with Pierson and the wonders he had accomplished here at Ramapo: "A fine example of prosperity which results from enterprize and industry. Mr. Pearson from the humblest beginnings has gradually risen to independence, if not affluence, still both he and his family still retain their habits of industry and frugality." Not all of Beck's observations made at Ramapo were connected with the works. He took particular notice, for example, of the rocky summit of nearby Mount Torne, a name which is derived from the Dutch word for steeple.

When the Erie Railroad came into the valley, the use of the four- and six-mule teams ended. For more than 40 years, heavy loads had been hauled over rough roads from the works to Haverstraw on the Hudson River. These "waggons" followed a much shorter route than the customary one south to Suffern and through Ladentown. Using an old wood road running east from the works around the base of Mount Torne, they proceeded northeast up the Torne Valley to Pine Meadow, turning east there to Ladentown and then on to Haverstraw.[2] A sloop, the *Josiah G. Pierson,* built expressly for the purpose, carried the iron down the river to New York City.

In 1851 Jeremiah Pierson, at eighty-six, leased the ironworks to Constable and Wilson, thereby terminating 56 years of continuous management for the Ramapo works. Three years later the dam at Ramapo was broken by a spring freshet and released the waterpower on which the works had depended for so many years. This event, followed by Jeremiah's death in 1855, marked the closing of the Ramapo ironworks and, with it, an era.

Some of the quaint old homes where the workmen once lived are still standing on the far side of the river. Several of the houses probably date back to the early days of the works at Ramapo and possess a charm of their own. However, the nail works, the rolling mill, and steel furnaces, which stood on the strip of land between the houses and the river, have long since disappeared.[3]

SOUTHFIELD IRONWORKS

When the Southfield ironworks started in 1804, it may have been intended as a branch of that famous neighbor to the south, the Sterling works.[1] The same Townsends who owned Sterling built it; many of the same names of customers are found in the ledgers of both works; and the old furnace at Sterling seems to have been abandoned the same year that Southfield came into existence.

The earliest entry in the Southfield records sets June 6, 1804, as the first date of activity there and by early summer in 1806 the furnace had been completed and was in blast. But running accounts of the operations of Southfield do not appear until 1810, when Peter Townsend II made the first blistered steel manufactured in New York State from iron ore of the Long mine at Sterling. Six years later he made the first cannon ever cast in the state from ore of the same mine. They were cast in 6-, 12-, 18-, 24-, and 30-pounder sizes. Not one failed when tested; some were light field pieces and all were manufactured for the U.S. Government. The 6- and 12-pounders, although modeled after brass field pieces of the British Army, were lighter in weight; yet the metal was fully able to withstand the continuous shock of firing, proof of the quality of the metal which was put into them.

The record books reveal that ironmasters of the Ramapos often did business with each other, a circumstance that might arouse suspicion if practiced today. For instance, entries from 1818 to 1820 in the ledger of the Southfield works report the following transactions: Abraham Dater bought "Pig mettle," stamped-iron, and scrap in great quantities. Southfield purchased from Dater share molds, "land sides," bars of iron, ladle molds, gudgeons, plates, and grates; and from Martin Ryerson, the proprietor of the Ringwood, Long

Southfield Ironworks, 1835. Oil painting by Raphael Hoyle. Size 46½″ × 34½″. The Manor House is at the extreme right; the furnace with its covered water-wheel and casting house at center right. Painting owned by the author. *Photo by Frank Moratz.*

Pond, and Pompton ironworks, Southfield bought share molds, drafts, and wagon boxes. The Southfield works sold Thomas Ward pig metal, stamp iron, scrap iron, cast plates, burnt plates, lumps of iron, and pig iron; Andrew Suffern "pig mettle" and stamped iron, and Ward and Dater 2,779 hoop poles.

Stamped-iron or "stamp iron" mentioned above was a major product of the Southfield works. The stamping mill was located next to the furnace and from the number of entries in the ledgers alluding to stamped iron, it must have been a very busy spot.

The varied tasks assigned to Daniel Green, an employee at Southfield, is an indication of the all-around type of work required of workmen:

> By 24 days Putting in Inner Walls and Harth in Augt. and Sept. 1818; By Keeping Furnace four months and twenty-five days in 1818; By Cutting 134 Cord Wood in the Winter of 1819; By 24 days Tending

Stamping Mill in April 1819 and 27 days the same in May and June; By Keeping Furnace five and one-half Months Blast of 1819; By Stamping fifteen tons last fall.

Another set of equally diverse duties were the lot of Samuel Conklin and his sons who also worked for the Townsends at Southfield—cutting wood, hauling rails and timber, running a sawmill, washing ore, hauling stone and coal, working on the turnpike, and even shearing sheep. The old ledgers are filled with a miscellany of details about life and labors at Southfield such as shipping iron bars from the Sterling anchory to the Southfield works; hauling timber in March of 1821 for construction of a new flume at the furnace; and various lists of employees such as David Ball, William Cable, William Colfax, Daniel Hall, Jacob Green, John Garrison, Nicholas Fennor, Peter Hennion, Daniel Morgan, and John Morgan. A description of an old coal wagon used at the furnace is given on a back page of one ledger. It was 12 feet in length, averaged 3 feet in height, with widths 5 feet at the top and 38 inches at the bottom. It could hold "one hundred and five bushels and one-tenth of a bushel." A record book kept by Peter Townsend III says that the ore carted to the furnace from October 20, 1834 to December 29, 1836 was taken principally from the Long mine, with occasional small amounts from the Patterson and Mountain mines. This particular book contains individual receipt stubs which are duplicates of those given to each driver showing the weight of the load, the mine from which the ore came, and the driver's name.[2] The loads averaged about one ton and each driver carted only a single load a day, as a rule. During the period covered by the book, over 364 tons of Long mine ore was hauled to the Southfield furnace.

In 1835, William H. Townsend commissioned Raphael Hoyle to paint a large landscape called "View of the Southfield Iron-works."[3] This painting was exhibited in the eleventh annual exhibition (1836) of the National Academy of Design in New York City, and it hung for years in the dining room of the ironmaster's home at Southfield. When the author bought it from a Townsend descendant in 1955, it was so obscured by age that identification was almost impossible. After being cleaned and restored, the canvas revealed the only existing picture of the ironworks and its surroundings as

No. 297

Southfield Furnace, March 21 1835

Delivered by George Patterson

team	T.	Cwt.	qrs.	lbs.
	1,	1,	2,	0

Long Mine **Ore.**

Detail from the previous picture. This shows an ore cart, and a teamster's receipt from the same year, 1835.

they appeared in their heyday. The furnace, casting house, gristmill, barns, stables, ore wagons with their drivers, and the ironmaster's house are all clearly shown. At this period the furnace was producing an average of 750 tons of iron annually, consuming 225 bushels of charcoal for every ton. Four years after Hoyle painted it, the furnace was dismantled and rebuilt.

While making a survey of the mineral district for New York State in September, 1836, Lewis C. Beck visited the Southfield furnace. At that time the main office of the Sterling Iron Company was located there and Beck was cordially invited to be the guest of the company and stay with Mr. Whitney, an agent. Beck's diary gives many previously unknown facts about the operations and management of the Southfield works. It says, for instance, they were preparing to smelt iron by use of anthracite, an undertaking which was promising, but still required larger scale tests to determine the real value of the process. Whitney told Beck that most of the preliminary tests had already been made by Pierson at the Ramapo works. Seven months later, on April 21, 1837, Beck paid the Southfield ironworks another visit and reported in his diary:

> The experiment of using anthracite in the smelting of iron at this Furnace appears to have been a failure, at least if iron can be manufactured in this way. The expense is greater than by the ordinary method. The difficulty here was that the ore, when mixed with the flux and charcoal, was not carried directly through the process, but after being heated was allowed to cool and then removed to another furnace. The material, of which the furnace for carbonization was constructed, was also an imperfect one and the flues were not capable of sustaining the intense heat which was applied to them.

Beck's notes confirm the experimentation that was being carried on in the early unsuccessful attempts to make pig iron by using anthracite rather than charcoal in the Ramapo region. The first successful New Jersey hard-coal furnace was built in Stanhope (1840–41) and had its first blast on April 5, 1841, about three years after Southfield made its unsuccessful attempts. This was a very important occasion, particularly for Pennsylvania with its rich anthracite mines, because the iron was of very high quality. The Southfield furnace was not converted to anthracite until 1868.

Meanwhile the charcoal method was still capable of keeping Southfield busy. In one blast from October 11, 1850 to July 3, 1853, they had made 6,353½ tons of iron even with 62 days stoppages. The Southfield furnace record books show several other rather long continuous blasts from June 10, 1859 to November 3, 1860; April 20, 1862 to March 12, 1863; and September 28, 1863 to March 4, 1865. When combined, these blasts produced an average of slightly more than 43 tons of iron per week. Ore was plentiful from 1859 through 1866 with the Augusta, Cook, Hard, Middle, New, and New Augusta mines furnishing various kinds, all of sound quality.

Excerpts from the Southfield furnace blast account records contain an amazing amount of data on daily events and equipment at that time. A memorandum describing the furnace's first blast in 1859 reads like an on-the-spot observation: "On Sat. June 5th, filled the Furnace and ignited the charcoal at 4 PM. Weds. June 9th, put in ore at 11 O.C. Furnace down 7 charges. The ore came into the Hearth in running 26 charges and 48 hours coming through. Made all necessary arrangements and commenced blowing, Friday 11th at 11 O.C. AM."[4]

Precise specifications of furnace dimensions and component parts, with exact amount of charges, coal, flux, and type of ore used are all painstakingly noted as guides to follow (or avoid) for the successful completion of future blasts. Some of the difficulties that halted production were written down, perhaps as alibis for not meeting pre-set schedules, or just as examples of the troubles that plagued the furnace operation.

> Stopped 4 hours, put in new dam plate and runner, lost one casting. . . . Back heating pipe gave out. Commenced filling blank charges. . . . Stopped, repaired waterwheel [24 foot overshot], lost 18 hours and one casting. . . . Stopped up the Furnace in consequence of running short of Charcoal.

On April 1, 1864, both the Southfield and Sterling ironworks were sold as a unit to the newly-formed Sterling Iron & Railway Company in which Peter Townsend retained an interest. In 1868, the Southfield ironworks converted to anthracite, believing that production would not only be increased, but also be more efficient. The panic of 1873, however, which adversely affected every ironworks in the

Southfield Furnace, shortly after it closed down in 1887.

nation, did not spare Southfield. However, activity picked up by 1878, and the ledgers show that the blasts from 1878 to 1887 were regular and sustained, including one of 117 weeks in which 12,788 tons of iron were turned out.

A very important and interesting phase of early iron manufacture was the preparation of the furnace prior to its being fired and going into blast. The Southfield furnace blast account book furnishes detailed descriptions of this activity:

> Commenced filling Southfield Furnace for Blast No. 10 on Thursday July 14, 1881, and put in three tiers of wood on end and seventeen tons of coal. Fired the Furnace on Monday July 18th at 9.40 AM. Wood burned out and Blast put on at 9.35 AM July 19th. Have nice run of cinder at 5PM July 19th and casted 11 PM July 19th. Iron white. Mr. Townsend fired the Furnace. Blast No. 11. Commenced filling up Southfield Furnace August 14, 1882 and the first day put in wood on end to top of Boshes. August 15th commenced filling Stock. Got filled up Wednesday morning and Mr. Barlow fired the Furnace 9:34 AM

> August 16th. Fire worked nicely and Mr. Townsend put on the blast 8:31 AM August 17th. Have no explosions of gas and gas came immediately to Ovens and Boilers. Had run of high cinder 5:45 PM, first cast 12 PM August 17th. White Iron. A. W. Humphreys, the President of the Company, came here on the evening of August 16th and stayed until the following afternoon, everything working to his satisfaction.

From 1880 to 1887, ore from the Scott, Sterling, Crawford, and Red Back mines, all part of the Sterling group, was regularly used at the furnace; some was also used from the Bering mine from August to November of 1881. Charcoal was first used for fuel, then anthracite, and in September, 1880, coke was introduced.

With the death of Peter Townsend III, on September 26, 1885, the long reign of ironmasters in this family ended and Southfield furnace closed down soon after. Townsend had been born May 13, 1803, and he spent his early childhood at the family homestead in Chester, New York, diagonally across the street from the historic Yelverton Inn where the boundaries of the Wawayanda and Cheesecocks Patents were determined in 1785. He was educated at schools in New York City, Newburgh, and Long Island, afterwards serving an apprenticeship in the counting-house of Jacob Barker of New York before becoming a merchant in Canandaigua, New York. There followed a period while he worked with his father and brother at the Southfield ironworks, before returning to Canandaigua to marry Caroline Parish, July 9, 1828.

Later he went to England to inspect a large ironworks to learn construction techniques before building furnaces and rolling mills. About 1848 he erected a large works at Brady's Bend, Pennsylvania, shortly before returning to Southfield.

A tall powerful man, he weighed over 240 pounds and measured 54 inches around the chest. His familiarity with every step of iron manufacture made him an expert in his field. A man of character and integrity, he was beloved by his workmen. It was said that Townsend, at Brady's Bend, took the tongs himself and pulled a bar of railroad iron from the roller. It was the first ever rolled in America.

The Southfield furnace had its last blast in September, 1887. Its crumbling ruins lie almost a mile west of Route 17, on the north side of the Old Orange Turnpike to Lake Mombasha and Monroe,

New York. The high double arches that connected the hillside with the top of the furnace still stand although the underlying bricks are disintegrating. Its interior lining, in good condition, can still be seen—a rare example of a blast furnace used during the earlier days. The lining, built of bricks, is symmetrically shaped like a huge hen egg standing on its larger end.

The pig beds and casting house were located at the eastern end of the furnace. Portions of the walls of this building are still standing. Stretching east from this spot is the abandoned road bed of the Southfield Branch Railroad once connecting with the Erie Railway in the village of Southfield. Construction of this mile-long spur line from the ironworks was started in 1868 and it was completed at a cost of $17,925. In 1871 an almost unbelievable total of 4,120 miles was covered on this short track by the line's only engine

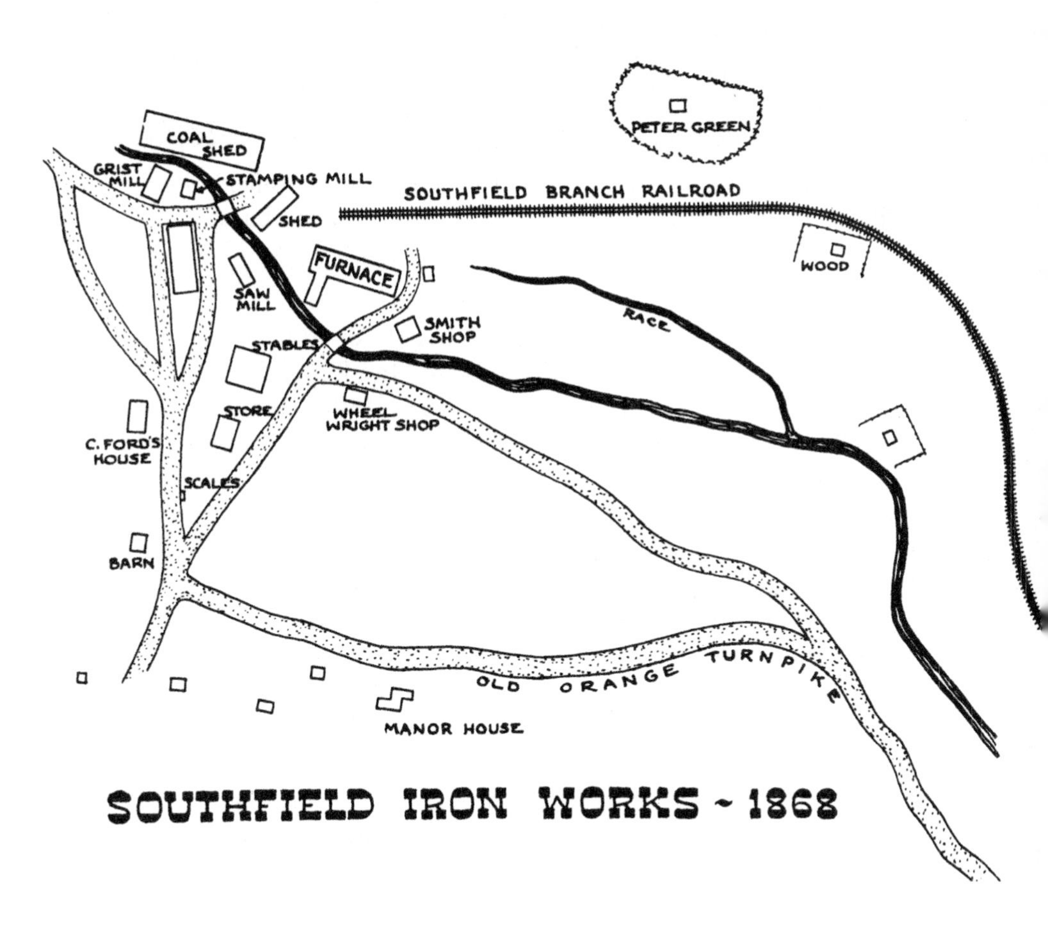

SOUTHFIELD IRON WORKS - 1868

Sterling Manor, the home of the Townsends at Southfield. *Photo by William F. Augustine, 1957.* Fire has since leveled the house.

which pulled 15,910 tons, consisting mostly of iron and iron products. Earnings for the same year amounted to $1,337.53.

In an interview during October of 1947, Josiah Jones of Suffern, New York, remarkably alert at ninety, recalled his experiences as a brakeman on this little railroad. He related one incident when the engine and two cars went over an embankment with such force that the sole of his shoe was torn off but, except for a bruised leg, he escaped unhurt. After Southfield furnace went out of blast permanently, Jones worked at Greenwood where he helped dismantle the works. The railroad line at Southfield was used seldom after the furnace shut down and by 1889 it was no longer profitable. Only 710 tons of freight were carried that year and earnings amounted to a mere $76.08.

In 1959, the City Investing Company, present owners of the Sterling Property, purchased the old ironmaster's mansion at Southfield, but it has since been destroyed by fire. Raphael Hoyle sat on a rise of ground behind this house when he painted his picture of the Southfield works in 1835. At one time there had been a fine library on iron mining and manufacturing within this home, as well as a number of valuable old paintings and relics of Southfield's past. Fortunately these were all sold before the fire. The author acquired a number of volumes and objects himself and the remainder went to other private collections.

STERLING IRONWORKS

The colorful history of the Sterling ironworks goes back to the year 1736. Cornelius Board (said to have been an agent for the Earl of Stirling) wandered up the Taughpomapack River [1] in the Ramapo Mountains to the mouth of a small pond, searching for

Sterling Furnace No. 3, prior to its closing in 1891. *Courtesy of Roscoe Smith.*

mineral deposits. The river later became known as the Ringwood and the pond as Sterling Lake. Board discovered outcroppings of iron ore at the southwest end of Sterling Lake. Today, the shafts of the Sterling and Lake mines lie here.

The location was ideal for producing iron. Nature had provided waterpower, a reservoir, virgin forests for timber and charcoal, and an apparent abundance of iron ore of a good grade. On October 1, 1736, Board obtained by warrant from John Burnet, an East Jersey Proprietor, a tract of 100 acres for himself and his partner Timothy Ward. On November 10, he acquired an additional 50 acres from Burnet. Soon after, Board and Ward erected a bloomery, where they produced the first iron made at Sterling and in 1738 they built a forge.[2] On February 13, 1739, they acquired from Burnet 10 small tracts of land in the Sterling area, ranging from ½–10 acres, totaling 27.72 acres. Presumably these tracts were bought to provide additional locations for waterpower and mining operations. All of these property acquisitions from Burnet were believed to have been within the limits of Bergen County, New Jersey, at that time.

The partition line between the provinces of New York and New Jersey, a subject of much controversy over the years, was not settled until 1769, when a commission appointed by George III of Great Britain determined its position. According to the new boundary, the Sterling ironworks was located in New York rather than in New Jersey.[3] Property owners, like those who owned the Sterling works, found themselves in an awkward situation. They sought to confirm their property rights. On September 26, 1772 the General Assembly of New Jersey passed an act upholding the titles of those who had purchased their land from the East Jersey Proprietors or from those whose land had originally been a proprietary grant.

Among all those property owners who were named in this New Jersey Act were: Cornelius Board, Timothy Ward, William Smith, William Hawxhurst, and James Burling, all of whom at one time or another were owners or part owners of the Sterling ironworks. By 1774, when all controversies had been settled, the commissioners marked the state line with large stones. Some of these boundary markers are still in place today.

In 1740 Cornelius Board sold out his interests at Sterling to his partner Timothy Ward and moved to Ringwood. Ward seems to have had a silent partner named Colton, with whom the purchase was made. On July 23, 1740, a total of 370.9 acres was surveyed to "Timothy Ward, William Smith and Company" in the Sterling area. It is possible that Ward was involved in two separate land ventures in the region at the same time. Two months earlier, on May 2, 1740, a deed in the nature of a quitclaim for 100 acres at Sterling was given by William Smith, James Alexander, and John Chambers to Colton and Ward. A little over a year later, Colton and Ward deeded the parcel to Edward Burling. Edward Burling then deeded to Samuel Burling an eighth part of this land with covenants on July 18, 1742, for 315 pounds.

The last four property transactions were among those whose titles were affirmed by an act of the New Jersey General Assembly passed on September 26, 1772.[4]

In the early days the name of the works was sometimes spelled "Stirling" but when the noted surveyor, Charles Clinton, was laying out the Cheesecocks Patent, between 1735 and 1749, he entered the name of the ironworks as "Sterling" rather than "Stirling." Clinton's field record, called the "Marble Book," because of its marbelized covers, was in the possession of the Townsend family until 1919, when MacGrane Coxe, who had married into the family, gave it to the Orange County planning board in Goshen, New York. The Cheesecocks Patent, granted by the Indians to a group of New York merchants in 1702 and confirmed by Queen Anne on March 25, 1707, covered a huge tract of land in the southern part of Orange County. After he surveyed the patent, Clinton, according to his instructions, divided the property into fair shares for each of the patentees.

Three tracts of land containing 131¼ acres at Sterling were resurveyed May 7, 1750 by James Burling. Ward and Colton are believed to have hired a stone mason, Peter Green, the following year to build the first furnace at Sterling. In 1752, the Nobles built a forge near the furnace, turning out their first anchors in 1753.[5]

William Hawxhurst's first connection with Sterling occurred on November 29, 1757, when he acquired almost 11 acres. On July 20, 1761, he added slightly more than 10 acres and on April 13, 1768, bought three more tracts comprising nearly 101 acres.

A few years after he had acquired property at Sterling, James

Burling died. Consequently, on October 20, 1755, this advertisement was inserted in *The New-York Mercury:*

> To be sold at public vendue, on the 18th day of next month at the Merchants-Coffee-House, the sale to begin at 11 o'clock, ⅛ and ⅔ d's of a ⅙th part, of Sterling Furnace, with ⅛ and ⅔ d's of a ⅙th part of the dwelling-house, coal-house, and about 600 acres of land, being the part belonging to the estate of the late James Burling, deceased. Said furnace and land lies in Bergen County, in the eastern division of New-Jersey: It is well supplied with oar, within a few roods of the furnace, and plenty of water in the driest season. Also to be sold at publick vendue, at the same time ⅞ths of a forge, called the New-Forge, with two fire-places in it, in good order, for making bar-iron, with about 100 acres of land belonging to it, and lies about 3 miles from the above premises, information may be had of John Burling in New-York, or Samuel Burling, in Burlington. Also to be sold, or let, on a lease for a term of years, by the said John and Samuel Burling, one eighth, and one twelfth part of the above said furnace and land. Octo. 18, 1755.

One of the chief sources of information about the Ramapo ironworks are these frequent, descriptive notices offering property or products for sale. Often they supply details concerning methods of transportation, geographical locations, natural resources, and other important data which would otherwise be unavailable. The Sterling works, particularly, appears over and over in the pages of these Colonial newspapers. A good indication of the market for iron, for example, can be gathered from this appeal in 1759 for employees of various kinds:

> Good Encouragement given by HAWXHURST and NOBLE, at Sterling Iron Works, for Wood-cutters, Colliers, Refiners of Pig and Drawers of Bar Iron. Also a Person well recommended for driving a four Horse-Stage between said Works and the landing. N. B. Pig and Bar Iron and Sundry English Goods to be sold by William Hawxhurst in New-York.[6]

Hawxhurst and Noble seem to have formed a partnership at Sterling some time during the two-year interval between 1757 and the date of this advertisement. In 1759, William Noble died and left his interest in the ironworks to his son Abel, who continued the partnership only briefly. On October 19, 1760, Hawxhurst announced that he would carry on the works alone.

One of the largest problems confronting Hawxhurst was transportation to New York City. He solved this in 1760 by petitioning the Provincial Assembly of the State of New York for the right to construct a road from Sterling across the county to Haverstraw on the Hudson River. There were others who had interests in the ironworks and mines, as well as persons inhabiting and holding lands in the county, who would benefit from such a road. The request was granted as it seemed reasonable for the success of the ironworks, and at the same time it could provide a short route to the river for those living behind the highlands. Hawxhurst, Charles Clinton, and Henry Wisner were named commissioners and were empowered to proceed, at the expense of the petitioners and any other persons who would voluntarily contribute, to lay out, clear, open, and complete a public road—not exceeding three rods in width—and on the shortest course from Sterling through the highlands to the most convenient landing place at Haverstraw. Charles Clinton was the engineer and surveyor for this road, which proved to be the benefit to Orange County that had been expected.

In May of 1761, Hawxhurst was unsuccessful in petitioning the Provincial Council of New York for the sole right to make anchors and anvils within the province. That year he advertised that he had added a refinery for Sterling pig iron, obviously believing in keeping the name of his ironworks before the public. This modern sales approach was necessary because of Sterling's location, isolated from prospective New York customers. He continued his campaign by frequent advertisements in the *New-York Gazette or Weekly Post-Boy* and *The New-York Mercury*. During July and August of 1764, Hawxhurst assured readers that he still carried on the Sterling Ironworks and sought:

> founders, miners, mineburners, pounders and furnace fillers, banksmen, and stock takers, [re]finers of pig and drawers of bar; smiths and anchor smiths, carpenters, colliers, wood cutters and common labourers. They will be paid ready cash for their labour, and will be supplied with provisions there, upon the best terms.

Nor did Hawxhurst neglect to remind the public that he

> continued to sell pig iron, bar iron and anchors, which he makes of any weight under 3500 (lbs.), and as he has by him a considerable

> amount of anchors, he would sell them by the ton, to retailers or exporters at a lower price than the importers from Europe, or the neighbouring colonies, he also casts waggons and chair tires, which he sells on the most reasonable terms, for cash or Connecticut Proclamation Bills. He also will take old and cast anchors in part payment for new ones, in proportion as they are in value.

Two years later, Hawxhurst put himself and his works into print again in the same papers with an announcement of having "lately erected a finery and great Hammer, for refining Sterling Pig Iron into Bar and takes this Method to acquaint his old customers, and others that they may by applying to him in New-York, be supplyed with flat and square Bar Iron, Cart, Waggon, Chair and Sleigh Tire-Mill Spindles; Wrines, Clanks, and Axeltrees, Cast Mill Rounds and Gudgeons." With an eye toward keeping old customers as well as attempting to interest new ones, Hawxhurst adds that he "continues to make Anchors as usual."

During this period of expansion at Sterling the actual owners remain a mystery. Abel Noble, as well as Hawxhurst, certainly had a financial part and probably owned the majority interest in the bloomeries, forges, the furnace, and the anchory. Just when Peter Townsend became associated with the Sterling ironworks has been disputed by writers for many years; previously, all have assigned too recent a date. According to a 1768–75 ledger, Townsend was a partner, with Abel Noble, as early as February 11, 1768. At this time Sterling was turning out wagon boxes, pig metal, tea kettles, skillets, pots, mill rounds, axel-trees, refined iron, and potash. It also produced in the same period 1,386 anchors weighing 185 tons, averaging about 268 pounds each; and 15,231 tons of bar iron. Sterling products were sent to such destinations as the Carolinas, Boston, Quebec, England, "Passagoula" on the Gulf Coast, the British West Indies, and Rhode Island. The ledger containing these entries prefaced shipments to these places with "Voyage to" or "Adventure to," an indication of the lengthy and uncertain nature of Colonial transportation. Many consignments were made to "David Anderson, Boteman, of New, say Old Bridge, N.J.," owner of a sloop used for shipping iron.

On May 23, 1767, the Sterling anchory was destroyed by fire. In October of 1768, an advertisement appeared announcing that

Sell for Account of. Noble & Townsend C[o]

1768	To whom Sold &c.	Number of anchors	Weight of anchors	Weight of refined Iron	Price		£	s	d
February	By Sundries	3	170		@ 9		17	18	6
11	1, By Cash recived for			0..1..11	37/	6	..	19	..
15	1, By Cash received from	1	216	abated 9	9	6	8	..	6
18	1, By Cash received for			1..-..1	39/	6	1	13	5
Ditto	2, By Cash as pr. Journal			5..2..9	do	6	8	16	7
March 3	2, By Cash received for	1	138		8½	6	5	..	7½
1	2, By Cash received for			12..0..21	28/	6	17	12	6
11	2, By Cash received for			1..3..13	30/	6	2	16	
12	3, By Cash received for			8..1..26		6	12	10	6
15	3, By Cash received for			4..1..18		6	6	18	7
18	3, By Sundries pr. Journal			27..0..23			31	7	4
21	4, By Sundries pr. Journal			4:3:8			7	11	10
22	4, By Cash pr. Journal			21:2:3		6	30	10	1
23	5, By Cash received	1	209	27:1:4		6	45	17	0½
-	By Peter Townsend & William [illegible]		123	30..1..14	@ 28/		42	11	6
	6, By Cash as pr. Journal	1	444	5..1..11		6	22	5	0½
30	6, By Cash received for			1		6	1	14	..
31	6, By Sundries as pr. Journal	2	480	35..3..24			68	1	0½
April 2	7, By Cash as pr. Journal			6:0:7		6	8	17	9
1	7, By Lambert Garrison	1	166		9		6	4	6
5	7, By Cash received as pr. Journal			6..2..4		6	11	3	3
8	8, By Cash received pr. Journal			3..3..25		6	6	..	8
9	8, By Cash received for do			8..0..21		6	12	13	1
11	8, By Cash received for do			4..3..23		6	8	1	4
12	9, By Cash pr. Journal			20..1..1	[illegible]	6	29		9
13	9, By Sundries pr. Journal (2 Entries)			21..3..17		6	32	13	3
15	9, By Cash pr. Journal	1	78	4..0..3		6	4	11	4
16	10, By Cash pr. Journal			9..0..22		6	13	9	2
18	10, By Cash pr. Journal			3..2..26		6	5	7	1
20	10, By Cash pr. Journal (2 Entries)			5..0..15		6	7	13	3
22	11, By Sundries pr. Journal			4..2..24	[illegible]	6	7	16	11
23	11, By Cash received pr. Journal			3..0..15		6	4	18	6
26	12, By Cash pr. Journal			23..3..9		6	32	1	8
27	12, By Cash pr. Journal	1	82	0..0..26		6	3	8	9
29	12, By Cash pr. Journal	1	68	103..1..2		6	137	2	10
May 2	13, By Cash received for	1	15		9	6	1	13	9
3	13, By Cash do for	1	117		8	6	4	7	
4	13, By Cash do for	1	71		8	6	2	13	
7	13, By Sundries as pr. Journal	1	89	4..1..26	[illegible]		10	7	
10	14, By Cash pr. Journal			10..0..13		6	16		
11	15, By Watson & Murray for			11..3..15	28/	[illegible]	20	16	

A ledger page, proving that Noble and Townsend were partners as early as 1768. *Courtesy of the American Iron and Steel Institute.*

it had been rebuilt and notified readers that Noble and Townsend were carrying on their anchory business, selling anchors of the best quality, costing half a penny per pound less than new ones imported from Europe or the neighboring Colonies.

Another ledger of this famous old works found at Sterling Manor contained entries of business transactions with Robert Earskin (Erskine), John Suffern, Andrew Tobon, Henry Townsend, Joseph Remington, William Genn, and William Fitzgerald from 1772 through 1774. It reveals that 385 tons of pig iron were produced at Sterling furnace from January 1, 1772 to December 31, 1773. Scrap iron and stamped iron were also sold in quantity. One entry in 1773 is marked "Hawxhurst to Furnace Account"; another, for 358 "Tuns Pigs" is billed to their store at Chester, New York.

The question of ownership of that part of the Sterling tract, originally bought by James Burling and advertised for sale in 1755 by John and Samuel Burling (presumably his sons, but possibly his brothers), was evidently in dispute. Twice in February and again in March of 1775, John William Smith and Samuel Burling offered "three-fourths of the STERLING FURNACE with Improvements thereon" for lease "from the first Day of April next." They claimed they also would lease "the whole of the Sterling Forge within three Miles of the Furnace, now in possession of Abel Noble, at which Place the famous Anchor Works have been carried on for many Years past." They believed that "further Description was unnecessary" because "those Iron Works are well known to exceed any on the Continent, both for their Improvements, and every other Advantage, as well as the superior Quality of the Iron." As additional inducement, a final note suggested that "Whoever is inclinable to Lease the above-mentioned Works, may probably have an Opportunity of purchasing or leasing the remaining Part of the Furnace, of the present Tenant."

This advertisement seems to have aroused Peter Townsend. On February 27, 1775, the following reply appeared in a New York newspaper:

To the Public

Whereas John W. Smith and Samuel Burling, have advertised to be leased three-fourths of the Sterling Furnace, with the improvements thereon, also the whole of Sterling Forge; I think it my duty to inform

> the public, that I am the proprietor of one-quarter of the Forge and Anchor Works, and all of the Furnace and the New Mine, with the meadows and building there-unto belonging, and that I will not dispose or give possession of the same to any person whatever. I am induced to give this notice to prevent any person being inadvertently brought into trouble and disputes by taking a lease for the premises, under the said John W. Smith and Samuel Burling.

Despite this claim and counterclaim, Noble and Townsend had resolved their differences with Burling and Smith to the extent of planning to enlarge the two forges available for making steel in 1775. In the *New York Journal,* December 28, 1775, Abel Noble and Peter Townsend announced "A FORGE with Six Fires to be built near Sterling." This new undertaking required "Anyone that inclines to undertake it, must give in their proposals before the 25 of January, as immediately after that it is intended to set about cutting and drawing the timber, as the roughness of the country makes it necessary to collect the timber while the snow is on the ground." Interested parties were directed to apply to William Hawxhurst in New York. This new forge was completed the following year and a call made for workmen in April. Whether this expansion was made with an eye on the impending Revolution is impossible to say, but the business sense of both Noble and Townsend makes it a possibility.

Soon Sterling was busy supplying the Continental Army with products. Filling these military contracts, however, became more and more difficult because of the manpower shortage. On August 8, 1776, this "Memorial of Abel Noble and Peter Townsend, Proprietors of Sterling Iron Works, Anchor Works & Company by William Hawxhurst, their Agent Humbly Sheweth" was presented to the Honorable House of Convention of the Representatives of the State of New York:

> That your memorialists have made a Contract for making Anchors, Steel and broad Bar, and large square Iron to a Considerable amount for the Continental Service which were to be made with all possible expedition. That your memorialists have already constructed a new work for wraughting the said Anchors and have made a considerable Progress therein. But unfortunately, for your memorialists, the Men of War's arrival up the North River have occasioned the county to raise

the Militia whereby the workmen and Labourers are taken from the said works.

That your memorialists had just before the arrival of said Men of War lodged at Cave's Store at Haverstraw Landing fifty eight Barrels of Pork, one hundred Bushels of Pease, five Hogsheads of Tobacco and sundry other necessaries for the use of the people employed at the said works of which said Provisions . . . the Militia have taken forty Barrels of Pork, seventy-five Bushels of Pease and three Hogsheads of Tobacco by means of all which and the loss of Time already sustained your memorialists are unable to perform their Contract this season unless your Honours shall give them Relief, by discharging the workmen and Labourers from said Militia, and supplying them with the Like Quantity of Pork, Pease and Tobacco as have been taken from them as foresaid and also with ten or twelve Sledge men which your memorialists humbly pray your Honors will do as speedily as possible.

Wm. HAWXHURST [7]

Two weeks later, August 22, 1776, Noble and Townsend filed another petition with the same governing body, emphasizing their desperate need of competent men in order to carry out the contracts for the Continental Army.

Townsend and Noble built a new anchor works, but because the laborers were still subject to draft they could not deliver the ordered materials as promised. They again asked that their workers be protected from the draft.

Reaction of the legislature to these pleas was slow and eight months later, Noble and Townsend were still having trouble keeping the necessary workmen for the furnace, anchory, and forges. On April 12, 1777, they once more sent a petition to the Convention of New York listing the following minimum requirements as bare essentials:

For the Furnace
20 Men, Wood cutters,
4 Master Colers,
16 helpers [for the colers]
3 Men for Raising Oar
2 Men for Carting Oar
7 Men Carters for Hauling Coles
2 Men for Stocking Coles

1	Banks man	
2	Men Burning Oar	68 Men
2	Mine Pounders	
2	Fillers of furnace	
2	Founders	
1	Gutterman	
1	Blacksmith	
1	Carpenter	
1	Manager	
1	Clark	

For the Forge and Anchory

20	Men for Cutting Wood	
3	Master Colers,	
12	helpers [for the colers]	
5	Men Carters for Hauling Coles	
2	Stocker of Coles	
10	Men for making Iron in five fires	66 Men
10	Men for making Anchors, three fires	
1	Carpenter	
1	Blacksmith	
1	Manager	
1	Clark	

For the Steel Works and Forge

15	Men for Cutting Wood	
3	Master Colers,	
12	helpers [for the colers]	
4	Men Carters for Bringing the Coles	
1	Stocker for Coles	
1	Man to Cart Pigs	49 Men
6	Men for making Steel, in three fires	
4	Men for making iron in two fires	
1	Carpenter	
1	Black smith	
1	Man to manage the Business [8]	

Despite the army work and claims of being shorthanded Sterling still advertised its iron products. It seems surprising that the following appeared in the *New-York Packet & the American Advertiser* on May 8, 1777:

> For Sale, at Chester, in Orange County, 17 miles from New-Windsor, on the North-River, opposite Fish Kill Landing, at the Sterling Iron Works. Sterling and Jersey refined Iron; likewise Steel, manufactured in the German way, from the Sterling pig, warranted to be good, that is, if it does not prove so, on trial, they will take it back and return the purchasers their money. Iron and Steel of the same make and quality has been formerly advertised for sale in New-York, by William Hawxhurst.
>
> Pork in barrel, live fat hogs, sheeps wool, good clean drest flax, New England Tow-cloth, white and checked flannel, and woolsey, and home spun woolen cloth, will be taken in payment. Noble & Townsend, Iron-masters at Sterling.

A postscript was added that has misled many historians into thinking it pertained to a second blast furnace at Sterling: "N.B. The Sterling Company are building a furnace for manufacturing Blistered Steel, which will soon be completed." A steel furnace is a much smaller structure than a blast furnace and entirely different in its appearance and function. In records now available, only one blast furnace is indicated there during the Revolution.

On February 2, 1778, Peter Townsend signed an agreement with the Continental Army to produce an iron chain which was to be placed across the Hudson River at West Point and act as a barrier for British ships.[9] As it turned out, the British never tried to get past the chain which was put in place on April 30, 1778.

They had their lighter moments at Sterling too. One hot June day in 1779, Aaron Burr arrived at Sterling with a winded horse and no fresh horse could be found immediately. After some time, a half-broken mule called Independence was procured and Colonel Burr mounted. The balky animal fully lived up to his name: he bolted with his rider and ascended a steep bank, where a little building, open at the top, stood filled with coal. Full speed ahead Independence, trying to shed his burden, dove headlong into the coal house. A chute of coal (reputedly thirty feet high) spewed forth. Mule and rider, still clinging determinedly, slid down with it in a cloud of coal dust. Burr unhurt, but scarcely immaculate, dismounted and persuaded a workman to lead the animal a mile or so before he remounted, and his journey was continued without further incident. It amused Burr to recount this anecdote to friends and younger members of the Townsend family.[10]

Nothing important can be found about activity at the Sterling works during the remainder of the Revolution. Peter Townsend died in 1783, leaving the property to his wife Hannah (Hawxhurst's daughter), who ran the ironworks with the help of their two sons Peter and Isaac, until Peter became older and experienced enough to take over the management himself. Peter Townsend II was born in 1770 and by 1783, showing his spirit, he rode horseback to New York to watch the British evacuate the city. As an ironmaster, he became a worthy successor to his father at Sterling.

The extensive tour of Sir Hector St. John de Crèvecoeur to the Ramapo Mountain ironworks after the Revolution included a visit to Sterling:

> We left Schunnemunk [Mountain] for Sterling, whose heavy hammers we were not long in hearing, and we arrived there early. No sooner had we put our horses in the stable, than the proprietor, Mr. Townsend, presented himself and received us with the politeness of a man often accustomed to seeing travelers and strangers. In other words, his hospitality is so well known and has been for so long a time that,[11] if one comes from the interior or from New York City, one always arranges to stay with him in crossing the mountains.
>
> Having learned that the object of our journey was to examine carefully the different works, he offered to show us all the details. First he conducted us to his large furnace where the minerals are melted and then converted into bars of pig iron weighing from sixty to one hundred pounds each. It was situated a slight distance away from the principal dam, which by the favorable position of the rocks made possible a reservoir of water of immense capacity. From a mere river, it was changed into a little lake of 15,000 acres in area,[12] filled with fish and on which he had a pretty boat.
>
> Draft for the furnace was furnished by two large blowers, 48 feet by 7,[13] which were made of wood only, containing no iron or copper. The violence, the noise of the wind which they produced resembled that of a tempest. This furnace, he told us, produces annually, when there are no accidents, from 2000 to 2400 tons of iron of which three-quarters are converted into bars and the rest melted into bullets, cannons, etc., for commercial use. These mountains, the wood from which furnishes me with charcoal, furnish also several kinds of minerals of an excellent quality known by different names.
>
> From there he showed us the refinery; six large hammers were engaged in forging bars of iron and some anchors, also several other

pieces for maritime use. Further down on the same stream was the foundry with a reverberatory oven. He showed us several ingenious machines designed for different uses for which they had sent him models which he had melted with a pinchbeck (an alloy of copper and zinc) recently discovered in these mountains, in which the grain, after the two fusions, acquired the fineness and nearly the color of tin. . . .

He told us how he is making three new kinds of ploughs, all lighter than the old ones and how he makes a portable mill designed to separate the grain from the small straws. . . . All this has left for Mount Vernon for, although General Washington is filling with his so distinguished talents the Presidency of the Union. . . . he still looks after the immense tillage and operation of his estate and farm, despite his duties elsewhere.

From the forge we went to see the furnaces in which the iron was converted into steel. It is not yet as fine as that from Sweden, Mr. Townsend told us, but we are getting there. A few more years of experience and we shall arrive at perfection. The iron which comes from my hammers has been known for a long time to be of good reputation, and it sells at from 28 to 30 pounds per ton of 2000 lbs. . . . I had trouble trying to cut down the brambles and thickets that cover a good part of the area here and finally solved it by obtaining 300 goats. . . . After having spent two days in examining these diverse constructions and admiring the art with which they had united the movement of the water, as well as the order and the arrangement of the logs necessary for furnishing the charcoal, which such a large enterprise requires, we left Mr. Townsend and the same day arrived at Ringwood, where we knew that the proprietor, Mr. Erskine had spent three years in Europe visiting the principal forges of Scotland, Sweden and Germany.[14]

Additional details and dimensions of the furnace at Sterling at this time are furnished by an old ledger kept there during the 1790's.[15]

Abel Noble, senior proprietor and part owner of the works, sold his interest to Peter Townsend II, who was representing Hannah Townsend and the company, June 5, 1797 for 1,800 pounds sterling.[16] Noble then moved to Bellvale, New York, where he died in 1806.

A tablet placed on the front of the Sterling furnace ruins in 1906 commemorating the West Point chain does not credit Noble. At that time—and for many years—the firm's full name was Noble, Townsend and Company but only Peter Townsend is mentioned, unfortunately.[17]

This old furnace seems to have been used for the last time about 1804 when the Townsends began construction at Southfield about six miles away. The old Sterling furnace was reputed to be in ruins by 1808, but the forges and anchory were still active, as well as a part called the Eagle Steel Furnace.

Most of its pig iron and forge products were sold at the company store in Chester, New York. Two-year ledger entries from the anchory and forges, beginning in mid-1813, list interesting customer names, nearly all from the New York and New Jersey area. "Blistered Steel and Crawley Steel delivered to Joseph Board. . . . Sterling Refined Iron sent to the New Cornwall Cotton Manufactory [Townsend was director]. . . . Iron sold to J. G. Pierson & Brothers [Ramapo works] and carted to Haverstraw Landing." There are entries for hauling horse feed from Earl's Mill and flour from Bull's Mill, as well as payment to Blackwell and McFarlan, at "Bellvale and Munroe Turnpike," for shoeing oxen.

Entries in the same ledger also carry those of the saw works starting from 1816. The saw works was erected by McCoun and Daniel and William Jackson during the War of 1812 on the outlet of Tuxedo Pond, two miles south of Monroe works, where Tuxedo Creek enters the Ramapo River. Originally this ironworks was built with the sole intention of producing mill saws, but the owners soon expanded it to include nails, plough shares, horseshoes, steel, forge bellows, anvils, whiffle trees, and a variety of other useful items. McCoun and the Jacksons had a part interest in the Sterling ironworks, which explains why the accounts of the saw works are carried in a ledger that starts with entries for the Sterling works. In fact there is written on the verso of the first leaf—in a fine bold hand—"McCoun & Jackson's, July 24, 1813, Sterling Ironworks."

In early 1814 the advantages of incorporating the Sterling ironworks became evident to the owners and they petitioned the New York Legislature. An act was passed on April 1 that brought into existence "The Sterling Company, Incorporated." It confirmed the interest McCoun and the Jacksons had in the works.

> Whereas Peter Townsend, Isaiah Townsend, Daniel Jackson, Henry M'Coan and others by their petition to the legislature, set forth that

> they are extensively engaged in the manufacture of steel and iron and that the vast and increasing demand for the manufacture of iron and steel is such that their private capital is insufficient for the purpose and that many persons are disposed to contribute moderate sums towards the extension of this branch of manufacturing. . . . Capital Stock shall not exceed $500,000.

About the time that the Sterling Company was incorporated, another enterprise was very much in the mind of Peter Townsend, for he bought a piece of property in New Windsor, New York, northeast of Sterling on Quassaic Creek, a tributary of the Hudson. Here, just west of a mill owned by Jacob Shultz, from whom Townsend purchased his land, he erected a cannon foundry. This consisted of two furnaces and four boring mills. In July, 1817, the first cannon ever cast in New York State was tested with great success. It set new standards in superiority of metal and in accuracy of firing. Despite this triumph, the cannon foundry was not a financial success and it apparently was taken over by the federal government. About 1836 John A. Tompkins obtained possession and converted it into a machine shop.

A gap appears in the records of Sterling between 1817 and 1825. In the latter year, the capital stock of the Sterling Company, Incorporated, was increased from a $500,000 limit to $750,000 and a pamphlet, *Report of the Committee Appointed to Examine the Condition of the Sterling Works*, was published.[18] This booklet, if it could be located, would probably fill in details of operations during these years at Sterling. Available ledgers have no entries for Sterling, but concentrate on Southfield furnace and works. There may have been some kind of litigation or financial difficulty which brought about at least a partial suspension of activities at the old ironworks in the Ramapos.

In 1847 the Townsends had started building another blast furnace at Sterling, which was completed and worked in 1848. It was just below the outlet of Sterling Pond, about 2½-miles downstream from the ruins of the old furnace at the foot of Sterling Lake. This new furnace had an inside height of 48 feet and was 13 feet wide across the bosh. It was converted from charcoal to anthracite in 1865–66. During a 48-week period in 1857, it produced 2,250 tons of iron from the magnetic ores furnished by the Lower

California, Upper California, Summit, Sterling, Crossway, and Mountain mines. A little community soon sprung up around the furnace area consisting of a church, school house, homes, a store, and office, and the usual workshops and buildings connected with an ironworks. For a number of years, church services were held in the school house.

The Townsends decided to sell their Sterling "iron estate" in 1856. A nine-page booklet was printed and distributed to selected men in the iron industry, in an attempt to find a buyer. It covered the mineral resources of the Ramapo Mountains and their advantages over those of other regions in lower costs of production, attractive market facilities, and the superior quality of Sterling ores.

The Sterling Tract is described as 23,000 acres:

> which is doubtless the most extensive Mineral Tract in the United States and that for nearly two centuries it has been occupied by the present proprietors and their antecedents as manufacturers of Iron, and on it is the oldest Iron Manufacturing establishment in America.[19] The Great Iron Chain of Revolutionary days was made at Sterling, as were all the Anchors and heavy Iron-work required for the first frigates and vessels of war that were built by the American Government after she became an independent nation. . . . it is a long established fact, that for castings requiring greatest strength, as Car Wheels, Engine and Ordinance Work, Malleable Castings and for Wrought Iron, its equal has not been found. Sterling Ores can be obtained in greater quantity and at a cheaper cost for mining, than from any Iron locality in this country; and it is confidently asserted and believed that the Ores can be delivered upon the banks of the Hudson River, within twenty-five miles of the City of New York, for one dollar and ten cents per ton, and at that price for a ton of pure black Magnetic Oxyde of Iron that will yield sixty-five percent Metallic Iron, which clearly establishes the very remarkable fact, that the Ore required to make one ton of Pig Iron can be furnished for less than two dollars, at a point on the Hudson River, where the best Pennsylvania Anthracite Coal can be delivered afloat, at about the price the same sells for in the City of New York; and where the iron, when manufactured, would be at market without further charge of transportation.

In their enthusiastic appraisal of the Sterling properties, "the proprietors of the Sterling Estate" invited "the most thorough

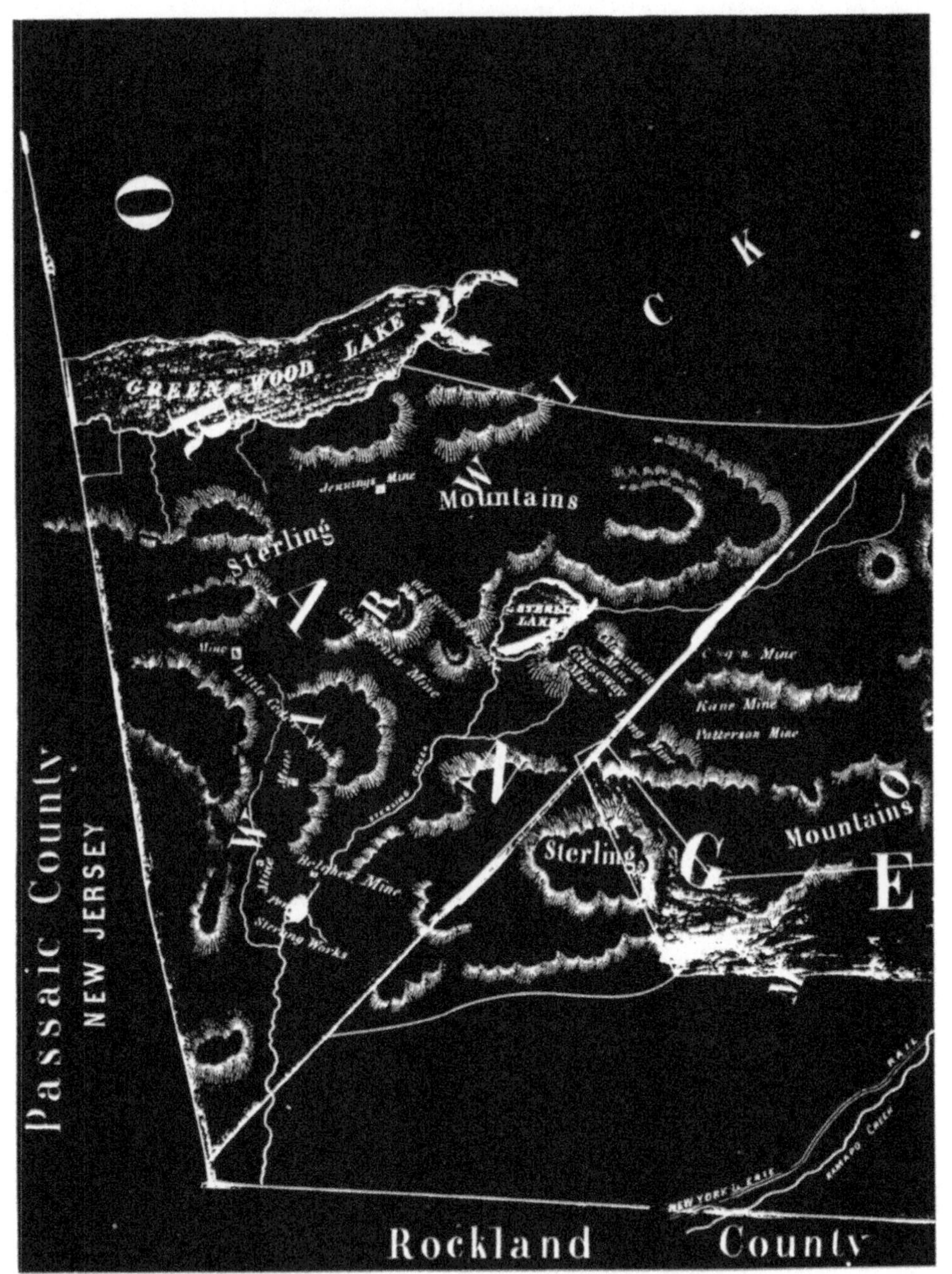

This map accompanied the booklet offering

the ironworks property for sale in 1856.

examination and closest scrutiny under a feeling of entire confidence, that their statements are substantially correct, and which they are well assured will be quite apparent to any person who would give a few hours to investigate and explore the estate, as millions of tons of the purest Ores are lying here and exposed to view."

With an eye toward the all-important factors of costs and profits, the pamphlet writer stressed "a very great advantage found in smelting these Ores [that] arises from their great richness and freedom from all baneful influences."

Considerable space was devoted to the economy and practicality of iron-making at Sterling. It cited the "cost of manufacturing one ton of Pig Iron at Sterling including delivery on the Hudson river is Calculated at $13.68." Projecting this tonnage cost and applying it to the "two hundred to two hundred and fifty tons of Pig Iron weekly . . . made from Sterling Ores without effort . . . it is positively certain that where Iron is manufactured as herein set forth, at a cost of $13.68 per ton, when commission and six months' interest is added to the cost, it will certainly show a cheaper production of Iron than can be exhibited at any other point." This statement bolstered the claim that the "net cash profit per ton" was figured at $10.24—an exceedingly attractive profit. Nor, added the pamphlet, was there any limit to the supply of ore and "business need only be circumscribed by the want of capital or the want of market demand for the manufactured article."

The pamphlet also compared the cost of the ores necessary for one ton of iron in favored localities in Pennsylvania, which was $9.25 per ton to $1.68 per ton for the Sterling ores mined on the banks of the Hudson River. Underlying the sober tone of the pamphlet one suspects a note of glee in the statement: "This immense difference in the cost of manufacturing Iron almost causes the proprietors to doubt the correctness of their own figures." It must have given them great satisfaction that their ore was much more economical than that of their rivals in Pennsylvania.

In contrast to the favorable financial aspects and advantages offered by the Sterling works, the physical description of the land and the equipment was somewhat limited:

There are on the Estate two large Charcoal Blast Furnaces, [at Southfield and Sterling] one of which is in blast, making nearly sixty tons of Pig Iron per week, of a quality that is not equalled, and commands a higher price than any Pig Iron made. Both furnaces are in perfect order, of modern construction with twenty-four feet overshot waterwheels, with Cast Iron Blast Machinery on durable streams.

The entire tract is thickly covered with thrifty wood and timber, in all of its various stages of growth, and which will prove an abundant supply for charcoal. There are, on the Estate, two fine rivers or streams for Hydraulic purposes, each of which has a large reservoir at its source, which makes them very durable, and the amount of Water Power on each stream is very great. The Estate is considered by the owners as indivisible and should be preserved in its present form, in which it presents attractions and advantages that are not found elsewhere.

Wm. Hawxhurst Townsend
Peter Townsend
Isaac Townsend
Geo. E. Townsend

P.S.—The Offices, Stores, Overseers' Houses, Workmen's Cottages, Barns and Stables are in very good order, and in sufficient number for the business.

There were no buyers. It was not until April 1, 1864, that the properties were sold. Through David Crawford, Jr., a son-in-law of Peter Townsend III, the whole of the Sterling works and properties were transferred to the "Sterling Iron and Railway Company." Peter Townsend III and others retained large interests in the property—among them were Thomas A. Scott, president of the Pennsylvania Railroad and assistant Secretary of War during the Civil War; Jay Cooke, who opened his own private banking house in Philadelphia in 1861 and was a very successful and highly regarded financier both privately and publicly; Joel Barlow Morehead; Samuel L. M. Barlow; and George C. Clark.

An interesting description of the Sterling works appeared in the *American Exchange and Review Magazine* in November of 1865:

In March of 1865, the furnace was blown out and converted into a hot-blast anthracite furnace, to be driven by steam, for which pur-

Sterlington Junction where the Sterling Mountain Railway joined the Erie, circa 1900. *Courtesy of Mary and Seely Ward.*

pose, a new engine of 150 horsepower has just been put up. All the works, excepting the outside stack of the furnace, have since been rebuilt in the most substantial manner. In a few weeks it is expected they will be relighted. It is not designed to make any alterations in the furnace at Southfield for the present, there being still an abundance of wood there for making charcoal. But around Sterling, for miles, there is scarcely a stick as thick as a man's arm, but has been scraped off and consumed in the insatiable maw of the furnace.

Mr. Chas. T. Ford has been Supt. of these Works for many years. The number of persons now employed in mining and repairing is about sixty; at Southfield seventy-five more are engaged in making charcoal or attending the furnace. Last year the entire force at work on the property, exclusive of railroad laborers, was between 175 and 200. . . .

The design of the present proprietors is not only to make pig iron on their premises, but to export large quantities of ore to Pennsylvania, shipping it to Piermont and thence by water to Philadelphia. The quantity of pig iron made in 1864, at the two furnaces, was a little over 3,500 tons.

The Sterling Iron and Railway Company immediately started to build a railroad for transportation of its products, from Lakeville at the foot of Sterling Lake, to a nearby point on the Erie Railroad about one mile north of Ramapo, New York, which they named Sterling Junction.[20]

> This road . . . is nearly complete and will probably be opened by the first of October [1865]. It is built in the most substantial manner. The gauge is similar to that of the Erie,[21] and locomotive power will be used in working it. At one point, known as the sink-hole, great difficulty was experienced in grading; and for weeks the earth dumped in disappeared, to rise up elsewhere. By the use of large quantities of brush, it is believed this difficulty has at length been overcome. The road terminates at Sterling hill, under a breast of ore hundreds of feet high, where it can be taken out in open day, at an expense of not over fifty cents per ton.

This new railroad was to be 7.6 miles long, with the highest grade 150 feet to the mile, favoring the direction in which the

An engine on a siding, circa 1900. *Courtesy of Mary and Seely Ward.*

heavy freight would be moving from the Sterling furnace and the mines, to its junction with the Erie.

By November of 1867, the rolling stock consisted of two engines, one passenger car and 142 freight cars. Fare for the passengers was three cents a mile. The cost of the road and equipment was $495,105. A separate company was formed to take charge of the railroad operation, the Sterling Mountain Railway Company, whose officers, according to a statement of November 30, 1867, were: J. Dutton Steele, Pres.; David Crawford, Jr., O. D. F. Grant, Peter Townsend, J. B. Moorhead, W. G. Moorhead, and Thomas A. Scott. The secretary and treasurer was A. W. Humphreys; the engineer and superintendent was Thomas C. Steele.

Comparison of figures on the operation of this local railroad for the years of 1867 and 1871 is listed as follows:

	1867	*1871*
Number of miles run by freight trains	17,040	8,250
Number of passengers carried	1,245	3,010
Number of miles traveled by passengers	8,702	15,050
Number of tons of freight carried in cars	93,227	64,000
Total movement of freight, tons carried one mile	644,838	410,128
Passenger earnings	$261.07	$603.45
Freight earnings	$53,972.39	$38,679.46

Several spur lines also started from the main track at Lakeville; one going east to the Scott and Cook mines, another traveling a very short distance north to the Sterling and Lake shafts. The two original locomotives, which were used for many years, were called Sterling Mountain No. 1 and No. 2. Sterling Mountain No. 2 locomotive was decorated with ornamental gold leaf and the headlight carried the Sterling coat-of-arms, consisting of a horse's head embellished with intricate scrolls. A single passenger car operated for many years, serving also as baggage car and caboose.

The history of Sterling after the Civil War is disconnected. Except for the names of some of the officers of the controlling company, little else can be found concerning the activities at the Sterling works until 1890. The Panic of 1873 had an adverse effect on the works here, as it did on all others. By 1880 high operating

Sterling Mountain Railway Engine No. 2.

During its heyday, Lakeville, western terminus of the Sterling Mountain Railway, at the foot of Sterling Lake.

Heavy-duty engine. All three pictures courtesy of the Railway and Locomotive Historical Society, Boston.

Sterling Furnace No. 3, circa 1898.

costs, overall market conditions, and the effect of the increased use of Lake Superior ores, which could be mined and transported at cheaper prices, took their toll on Sterling and other ironworks in the Ramapo region. One by one the mines and furnaces of the Ramapos disappeared and by 1890 the only furnace still in operation was the one built at Sterling in 1847–48 and even that one shut down finally in 1891, thus ending a century and a half of blast furnaces in the Ramapos. From then on all ore mined at Sterling and elsewhere in the region would be shipped to distant furnaces, mostly in Pennsylvania, where more modern plants provided greater profits and volume.

The assets of the company were dropping substantially, and in 1892 a reorganization of the Sterling Iron and Railway Company took place.[22] A year later only four of the 14 mines on the 22,000 acres were being worked and the number of miners employed at Sterling totaled 180. This gradual decline continued during the depression of 1892–96. Irregular use was made of only eight or nine mines, and from 1896 to 1917 only the Lake mine was really active at Sterling.

There were frequent changes in the controlling interests of the Sterling Iron and Railway Company during this period. MacGrane Coxe was president in 1892, then in a few years James D. Rowland of Philadelphia, who was succeeded by Theodore Price in 1905. H. A. Van Alstyne followed in 1911 and retained control until 1920.

Activity picked up at Sterling with the coming of World War I and the resulting demand for iron. About 1915 the Midvale Steel Company became interested in the mines. One division of this firm, the Ramapo Ore Company, Incorporated, through an agreement made with the Sterling Iron and Railway Company, was given the right to examine, explore, and test the mines and facilities of the Sterling property for a year from July 1, 1917 to July 1, 1918. Evidently satisfied with their findings, they leased the ironworks for a term of fifty years, commencing July 1, 1918. This lease arrangement, July 26, 1918, appears in a special 26-page

Sterling Furnace No. 3, which has completely vanished. *Photo by Vernon Royle, circa 1925.*

The office of the Ramapo Ore Company, now gone. Once located just east of the old charcoal furnace at the foot of Sterling Lake. *Photo by Charles F. Marschalek, 1952.*

booklet. The timing of this lease, however, turned out to be a poor one, for the signing of the Armistice in 1918, meant the end of profitable war orders. Despite dimmer sales prospects the Ramapo Ore Company spent large sums of money to recondition the mines and install new equipment, erecting among other things a $120,000 new, stone school and many homes for employees. They soon discovered their mistake. On July 1, 1923, after five years, the Sterling works were finally closed. For the first time, iron mining and manufacturing at Sterling, an operation that had been more or less continuous for a period of over 185 years, ceased. A four-month extension was granted for removal of the machinery and other equipment. With the departure of the superintendent, H. P. Sweeney, on November 1, 1923, a major phase in the history of this historic ironworks came to an end.

For approximately 29 years Sterling lay untouched, except for an incident in 1939, just prior to World War II, when the remaining rolling stock of the Sterling Mountain Railway was dismantled by acetylene torches and sold as scrap to Japan where it was undoubtedly turned into arms and ammunition—a sad end for the little railroad.[23] During World War II the U.S. Bureau of Mines and the U.S. Geological Survey explored and made field studies of iron deposits at Sterling as part of its wartime investigation of mineral reserves but the mines were not worked.

For most of the period between 1923 and 1952, the only human contact was the watchmen who acted as caretakers for the abandoned ironworks. Within the rugged and isolated hills of the Ramapo Mountains, few had occasion to visit or see the decaying works. Occasionally friends of the Sterling Iron and Railway Company owners were allowed to enter the area, usually for the purpose of fishing in the many ponds and streams that flow through the property. But Sterling had become a ghost works whose active days were over although its past was still evident in the shadows surrounding the mines, millraces, and piles of slag. Perhaps the

Lakeville. The building in the foreground, since razed, once held samples of drill cores. *Photo by Chester Steitz, circa 1935.*

most obvious and attractive feature that Sterling still had for a prospective purchaser was its rich history.

On September 18, 1952, the City Investing Company acquired the Sterling tract of 27 square miles from E. Roland and W. Averell Harriman. The property which they took over extended nearly nine miles north of the New Jersey state line, east of Greenwood Lake. It averaged almost three miles in width and cost $850,000. The final papers for transfer of the Sterling tract were signed in 1954.

City Investing Company's plans for Sterling and its vast acreage was wholly different from its former one of iron mining and manufacturing. The sequestered, relatively untouched, woodlands suggested creation of a flowering wilderness where no heavy industry would operate to pollute the air or violate the natural beauty. Outlining their plans for development, Robert W. Dowling and his associates projected an area to include, in addition to attractive administrative offices, a noteworthy group of floral gardens, the largest privately owned in the world, occupying 125 acres in all; a residential section of private homes; modern research laboratories; the cultivation of peat moss for private markets; a thinning of the woodlands by removal of slower-growing weaker trees and quarrying of rock, stone, and gravel from abundant out-croppings in Sterling.

Since 1954 many of these aims have been completed. The Sterling property has definitely become much more beautiful with man aiding nature. This can be seen especially at Sterling Forest gardens, four miles north of the furnace ruins. Colorful formal beds of flowers in the lush green setting attract visitors as do peacocks, flamingoes, and other exotic birds. Few realize that this was once an iron-making community.

Allan Keller describes the Sterling area as it is today in his column titled *Private Enterprise Forges Sterling Forest Paradise:*

> On the upper ridges of the Ramapo Mountains, just west of Tuxedo Park, lies Sterling Forest, which experts have called the best planned community of the future in the country. . . . The simple purpose behind this 30-mile square community is to provide a neighborhood for pleasant living, industrial research, academic inquiry, fun and recreation and woodland beauty. It's a real challenge, but my examination last week assures me it is going to come off. . . . The iron mines, the old smelting works and the forge are still there. . . .

About a decade ago it passed from the Harriman family to the City Investing Company. Robert Dowling, head of the firm, decided he would gamble on the concept that wisdom, intelligent land use and public-mindedness, combined with business acumen, could bring off a better success than any government agency with unlimited funds could achieve. Dowling and his aides knew that life in a heavily built-up area, with long commuting rides between home and work, wasted time, disturbed communal balance and provided no real surcease from the pressures of modern life.

What made life so pleasant in the small villages of our early days was the fact that so much was provided by the community itself. Dowling decided to attempt to bring back this communal balance with all of today's physical advantages. If you look at a map of the Forest you will see areas set aside for industrial research centers, another for NYU's woodland campus, a ski slope, public garden, lakes for boating and fishing, beaches on the lakes for residents and many scattered residential areas. But if you look at it on an aerial photograph, you will note the secret of future harmony. Industrial areas are separated from home sites by ridges. No building is allowed to rise above the tree line and winding roads and natural topography give complete isolation.

Already Union Carbide's nuclear research center, Reichhold Chemical center and International Nickel's experimental plant are functioning. . . . A large conference center has already housed meetings of the Ford Foundation. . . . Technicians, professors, scientists and others connected with these operations live in wooded seclusion, where deer browse in the backyards and the sound of ruffed grouse drumming in the woods is a sure sign of spring.

Anyone can buy a home. He need not have any connection with any of the institutions in the Forest. . . . But best of all, life itself will approach the level of quiet harmony our grandfathers knew 80 years ago.[24]

The scene at Lakeville, the old name for the community at the foot of Sterling Lake, is one example of the changes which most of the Sterling tract has undergone. Once a hustling and bustling mining center, with boiler shops, office buildings, power plants, carpenter and blacksmith shops, tool sheds, and railroad yard, modern Lakeville has reverted to its original state. A new road was constructed along a portion of the west shore of the lake so some of the old buildings connected with the works had to be razed. Even the old wooden church, formerly known as the Scott

Sterling Furnace No. 2, 1912. Thomas Edison is the third man from the left front, reading the commemorative plaque given by the Daughters of the Revolution. *Photo by Vernon Royle.*

Church, has completely vanished except for its rock foundation. This church had served the mining community for many years before heavy winter snow crushed the roof, leaving the building in such a dilapidated condition it had to be dismantled. The old office building just east of the furnace burned down about 1954 when the City Investing Company was taking over the property.

During World War I as many as five hundred men worked and lived at Sterling. So many of these employees were Italian, Polish, and Spanish that signs around the ironworks were written in foreign languages. Most of the miners' cabins are gone now—only three remain—but the foundations and fireplaces of others can still be seen. The Sterling Forest Corporation, a division of the City Investing Company, has built a very fine commodious lodge on the hillside at the southeast corner of Sterling Lake and other attractive homes and cottages for guests.

The furnace picture at Sterling has changed too. In the summer of 1955 the remains of the furnace built in 1847–48, near the outlet of Sterling Pond, were leveled by a contractor. Its stones were dumped along the sides of the spillway of Blue Lake which lies on the adjacent low ridge to the northwest. The City Investing Company had not yet acquired this part of the land. The older furnace at the foot of Sterling Lake, a mere heap of stones and dirt prior to 1958, is now restored. Work started in 1958 was completed in the summer of 1959, when a large plastic white dome, supported by huge white columns, was placed protectingly over it. Although done with the best of intentions, the dome was not in keeping either with the area or the furnace. Historians point out that the columns are more readily identified with Colonial Virginia than with the era of early iron-making in the rather wild, forested Ramapo Mountains. The dome has since been removed.

In the fall of 1959 Roland Robbins, well known for his restoration of America's first ironworks at Saugus, Massachusetts, was given the task of excavating the site around the old furnace. After only a few months' work he uncovered on the immediate north side of the restored stack, a waterwheel pit about 6 by 28 feet and some unusually thick walls. To the east of the pit he found the bellows room. Here he found two 50-pound weights—probably used to counterbalance the bellows. Another discovery made in this general area was a 50-pound piece of iron with the name "Sterling"

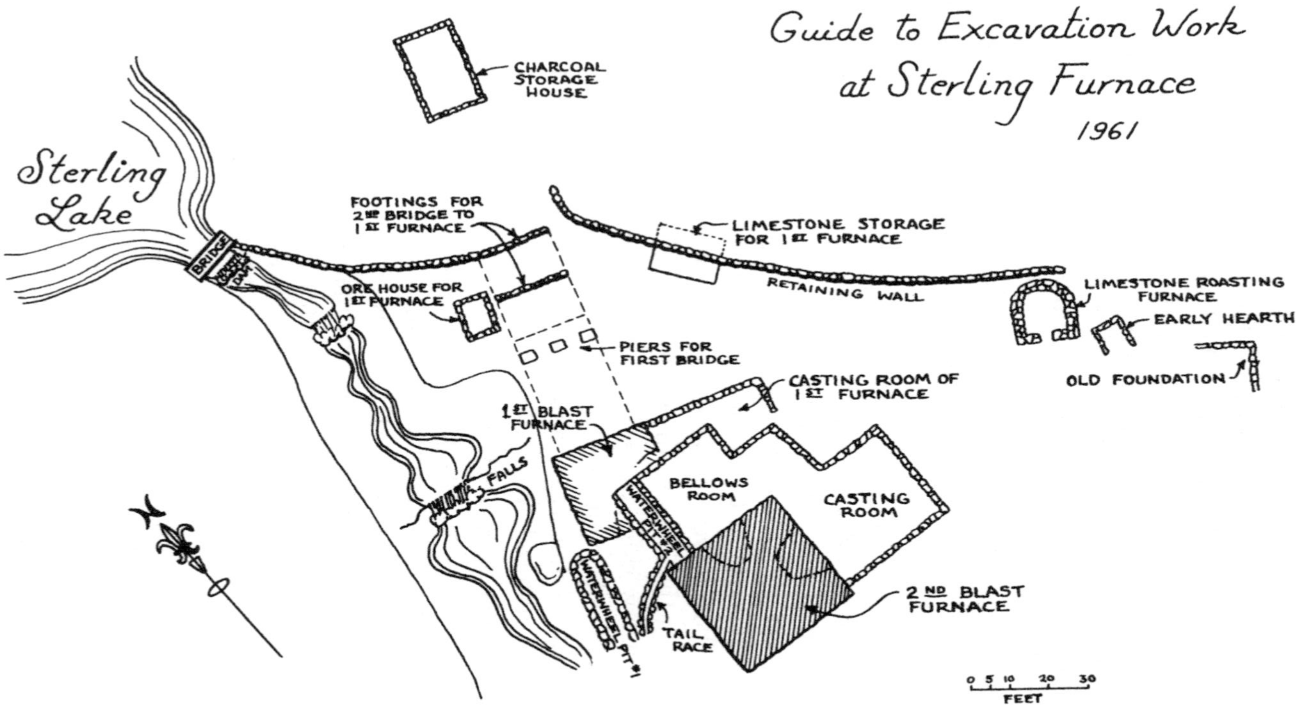

Guide to Excavation Work
at Sterling Furnace
1961
Sterling Lake
CHARCOAL STORAGE HOUSE
BRIDGE
FOOTINGS FOR 2ND BRIDGE TO 1ST FURNACE
ORE HOUSE FOR 1ST FURNACE
LIMESTONE STORAGE FOR 1ST FURNACE
RETAINING WALL
LIMESTONE ROASTING FURNACE
EARLY HEARTH
OLD FOUNDATION
PIERS FOR FIRST BRIDGE
CASTING ROOM OF 1ST FURNACE
1ST BLAST FURNACE
FALLS
BELLOWS ROOM
CASTING ROOM
WATERWHEEL PIT #2
WATERWHEEL PIT #1
TAIL RACE
2ND BLAST FURNACE
0 5 10 20 30
FEET

Excavation work by Roland Robbins at the old charcoal furnace at Sterling. Note the massive beams still intact in the waterwheel pit at Robbins' feet. *Photo by W. W. Hennessey.*

The old charcoal furnace, restored by the City Investing Company. The dome over the furnace has now been removed, but the columns remain. *Photo by W. W. Hennessey.*

embossed on it. Excavation of the adjacent hillside revealed a heavy deposit of charcoal and iron waste materials covering the original slope.

In uncovering the foundations for the casting house, to the east of the furnace, the location of a subterranean drainage system passing along the northern, eastern, and southern sides of the furnace was found. Several other foundations were also located but have been left for future excavations. The bed of the stream was considerably higher than when the furnace was in operation. All attempts to pump out the wheel pit proved futile because of the seepage from the stream. From this spade work it seems likely that the brook originally lapped against the western walls of the water-wheel pit and furnace. The excavations also clearly showed that many changes had taken place from time to time at the furnace. For instance, walls were found covering areas used for earlier iron-making.

In 1961, after a year's lapse, Robbins made a startling new find. Excavating operations uncovered a considerably older furnace just north of the previously restored one and revealed that the water-wheel pit of the more recent furnace had evidently been cut through a solid wall of its predecessor. But the discovery left unanswered the question of just when this older furnace was in operation and merely established the fact that there had been two different furnaces on the site at different times. The older furnace may have been the one built in 1751, but this leaves no date for the later one. Nor is the question resolved as to which one produced the iron for the famous West Point chain.

Robbins was also able to determine from his digging that the

Where lime was roasted, just east of the furnace. Note the iron pigs used as part of the equipment. *Photo by W. W. Hennessey.*

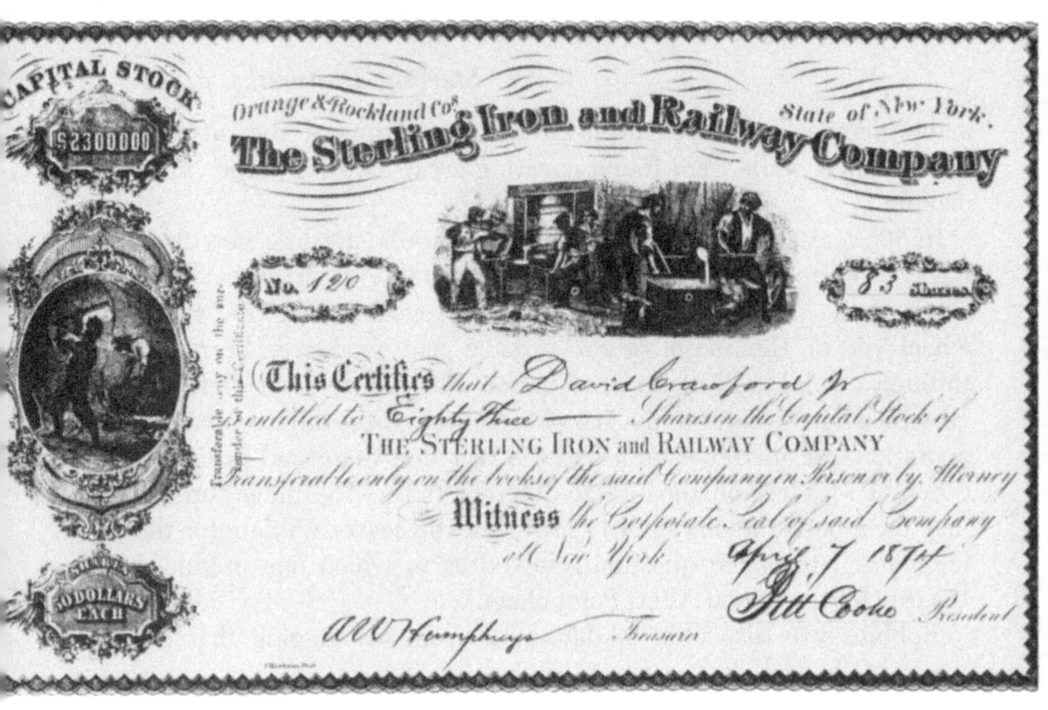
CAPITAL STOCK

$2300000

Orange & Rockland Co's State of New York.

The Sterling Iron and Railway Company

No. 120

83 Shares

This Certifies that David Crawford Jr

is entitled to Eighty Three —— Shares in the Capital Stock of

THE STERLING IRON and RAILWAY COMPANY

Transferable only on the books of the said Company in Person or by Attorney

Witness the Corporate Seal of said Company

at New York April 7 1874

Pitt Cooke President

A W Humphreys Treasurer

Transferable only on the surrender of this Certificate

SHARES 50 DOLLARS EACH

Stock certificate issued to David Crawford, Jr., in 1874.

older furnace was served by a rather oddly shaped waterwheel pit with round rather than the usual square ends. Below the water level of this pit, he found some original wood beams, used as part of the waterwheel foundation, with their joints and locking niches still plainly visible. To the east of the restored furnace, he unearthed the remains of a lime kiln with several pigs, used as a grate, still in place. The stream bed, lying just to the west of the furnaces and the southern side of the ridge where the outlet of Sterling Lake is located, was uncovered down to its original depth.

Not all of Sterling's past can be physically recreated today, but one needs only imagination while traveling through Sterling Forest to see the figures of pioneer ironmasters Board and Ward at work in their primitive bloomery, making the quiet forest ring with the sounds of their sweaty labor; to hear William Noble's voice raised to its highest pitch as he directs the work of building his dream forge near the furnace; or to conjure up six-mule hitches straining at their harnesses as they slowly pull heavy wagons loaded with anchors, pig and bar iron, over the dark and narrow roads through these Ramapo Mountain ridges on their way to the dock on the Hudson River at Haverstraw or to the company store at Chester.

JOHN SUFFERN AND HIS IRONWORKS

Of all the ironmasters in the Ramapos, John Suffern was one of the most interesting. He was born in northern Ireland near the town of Antrim on November 23, 1741, and came to Philadelphia when twenty-one, eager to take advantage of the opportunities to improve his fortune in the Colonies. Married in 1766, he and his bride, the former Mary Myers, moved to Orange County, New York, where he became a schoolmaster at Tappan in 1767. A brief period later they moved to Haverstraw and in 1773 settled permanently in an area known locally as "Point of the Mountain." He later gave the name of New Antrim to this section. After his death, it was named Suffern for him.

The house he built in 1775 was renowned for its size (documents record the length as 118 feet), and for its east and west wings. It served him both as a home and a public tavern which he operated for a number of years. The west wing, said to have been about twice the size of the east one, housed his family and contained two ovens. The east wing provided quarters for his slaves and employees. The earliest indication of an operating tavern seems to be an entry in John Suffern's ledger for the purchase of 127 gallons of rum from Samuel Skinner of New York on June 12, 1775. Earlier quantities, smaller in size, bought from Noble and Townsend at the nearby Sterling ironworks in 1773 and 1774, indicate these were probably obtained for his personal use, before opening the tavern.

The location of Suffern's tavern, near the junction of important roads leading north through the Clove to West Point; southwest to Ringwood, Pompton, and Morristown; and east to King's Ferry on the Hudson River, was undoubtedly the main reason it was so well known during the Revolution. As the Continental Army marched and countermarched across the southern New York–northern New Jersey section to meet threatened moves by British forces, Suffern's tavern became a familiar and favored stopping place for Washington and his officers. A southeast room, looking down the old post road to the New Jersey state line, is supposedly

Old John Suffern home, front and rear views, showing the smoke house and other buildings at Suffern, New York. Paintings by Jules Arnault, circa 1848; owned by Mrs. Jerome Cornell. *Photos by Louise Burnett, 1966.*

the one where Washington wrote letters during July of 1777 headed "Headquarters at the Clove." At the rear of the house, beneath a roof extended out over a back porch, Washington and his staff discussed and planned future moves of the Continental Army. When the French forces, en route to Yorktown, halted here August 19 and 20 in 1781, the Comte de Rochambeau, their commander, occupied a room in the tavern although his field tent was pitched a short distance away near the bridge crossing the Mahwah stream. An entry in John Suffern's ledger shows that seven cases of gin, bought from Charles White of Philadelphia in 1781, apparently arrived just in time for the use of the French officers.

John Suffern was a man of many interests. In addition to operating the tavern and his large farm, he bought a gristmill from Thomas Van Buskirk of New Hempstead, New York, on November 14, 1792. He also had a part in running a woolen mill; served for at least a short time in the commissary department of the Continental Army; represented Orange County as a member of the Assembly during the years 1781–82; and was postmaster of New Antrim from October 4, 1797 until 1808 when it was moved to the Ramapo ironworks, just up the road. A justice of the peace before becoming the first judge in newly-formed Rockland County in 1798, he also was superintendent of highways for Orange County in 1797 and later state senator from 1800 to 1802.

Although it was known that John Suffern owned and operated an ironworks after the Revolution, the record of its starting date came to light only in 1960:

> Article of Agreement made Between Andrew Suffern of the Town of Harvestraw and John Suffern of the Town of Hempsted in the County of Rockland and State of New York, their and each of their Heirs executors . . . the Twentieth day of February In the year of Our Lord one thousand eight hundred and eight.[1]

The plans of the Suffern partners, father and son, as outlined in this agreement included the building of a two-fire forge at New Antrim, another forge on "Minnersfaul Brook" at Haverstraw, a rolling and slitting mill as well as "several other Works such as Cuting and heading of Nails." [2] Just how many of these proposed projects were actually carried out is uncertain. Nor has any recorded data been found on the early years of the Suffern operations. But a ledger covering operations during the years 1824–1827 clearly indicates that a sizeable business was done at that time.

Entries in this ledger reveal that a considerable volume of miscellaneous iron products, such as ax, hoop, and bar iron, "crobars, waggon tire, square flat axel trees, landsides, shearmolds and scrap," was turned out at New Antrim. The customers of Suffern's New Antrim works had names that were not unfamiliar in the area: Peter S. Bush, James Townsend, Aron De Camp, Robert Knap, John Montague, and "John Goetschieus." Better known customers were Abram Dater, who owned two forges of his own farther up the Ramapo River at Sloatsburg; Gouverneur Kemble, owner of the West Point foundry and lessee of the Greenwood furnace at Arden as well as surrounding mines in the area; Peter Townsend, proprietor of both the famed Sterling ironworks and its neighbor, the Southfield works; Martin Ryerson who operated the Pompton, Long Pond, and Ringwood properties; and Jeremiah Pierson, proprietor of the nearby Ramapo works.

The names of employees Henry Whritenour and Antony Call are found often in the old ledger probably because of their ability to consume 166 pints of whiskey in a two-and-a-half month period. The ledger also records the wages paid by John Suffern as based upon the amount of iron the two men made. For example, "To H. Whritenour by making Iron with A. Call from Feb. 17th to May 27th, 1826, Tons 7–0–19, Pounds paid 56.3.0, your half of that amounts to 28.1.6 and for Flurrishing 0.9.6." Other workmen who are mentioned as being employed at New Antrim were James Thomson, Abm. Hammon, Phillip Man, John Still, and Richard D'Grow.

The inventory, taken after Andrew Suffern's death in October of 1827, gives a concise summary of the equipment with which the ironworks was provided. "1 pr. hammer tongues, 3 pr. hearth tongues, 2 pr. grampuses, 1 anvil, 2 hammers, 1 pr. forge bellows, 2 pr. bloom tongs, 1 pr. shingling tongs. . . ." John Suffern outlived his son Andrew and died November 11, 1836, within a few days of his ninety-fifth birthday.

James Suffern, a grandson of John, established a charcoal forge with two fires and two hammers on the Ramapo River about a half mile west of the former New Antrim, at Hillburn, New York, on the same site where his grandfather had built a sawmill in 1795. In 1852 he added a rolling mill to the water-driven works. James Suffern specialized in making rail-car axles until 1856. Three years later, Andrew Winter was the proprietor of these works which then consisted of the rolling mill with one puddling furnace to convert pig iron to wrought iron, two trains of rolls, and two hammers. About 300 tons of merchant bar iron was made out of scrap iron at the works in 1859. Winter employed about 25 men and almost the entire output was sold to the Erie Railroad whose tracks practically passed the doorstep. In 1872 operations ceased and the works were abandoned.

View of the second home built by John Suffern, located across the road from the present School of the Holy Child in Suffern. This looks west toward the Ramapo Pass with Hoevenkopf Mountain at the left. Painting by Jules Arnault, 1844; owned by Mrs. Jerome Cornell. *Photo by Louise Burnett, 1966.*

WOODBURY FURNACE

The Woodbury furnace, in Woodbury, New York, on the stream flowing from Hazzard's Pond, was built by Gouverneur Kemble in 1832. He also owned the West Point foundry and leased the Greenwood furnace with surrounding mines. In 1837, Peter Parrott was placed in charge of the Woodbury furnace and the following year took over the management of the Greenwood furnace for Kemble as well.[1]

Lewis C. Beck who visited the works at Woodbury on June 26, 1838, wrote in his diary about the heaps of ore from the O'Neill mine. This ore had to be roasted in a kiln before being used in the furnace. Sterling ore was also said to have been used there. On August 21, 1838, Beck and Peter Parrott visited the Greenwood (Patterson) mine, located "2½ miles nearly east from the Greenwood Furnace . . . on the Greenwood Tract."

In 1839, Peter Parrott and his brother Robert purchased the Greenwood ironworks from Gouverneur Kemble. Peter resigned as superintendent at Woodbury to take full charge at Greenwood, although Robert continued in his position of superintendent of the West Point foundry. The Woodbury furnace must have closed down very shortly after Peter's leaving, for in 1846, Samuel Eager writing about Orange County said: "Formerly there was a furnace in operation at Woodbury, but by the consumption of all necessary materials to conduct it for several miles around, the owners were compelled to let it go down and the establishment is in ruins."

Few details can be found about Woodbury. The furnace is believed to have been constructed principally of millstone grit, a reddish-colored native stone from nearby Pine Hill, New York. The set of hearth stones used in the furnace were cut in Cornelius De Pew's quarry, just north of New City, New York, by De Pew's grandson, who paid his grandfather $15 a set of stone blocks for a hearth. But beyond these few facts, Woodbury furnace is veiled in mystery. Like many of the furnaces in and near the Ramapos its story has become more legendary than factual from lack of records.

III

THE MINES

THE MINES

For many years the iron mines of the Ramapos and adjacent areas have aroused curiosity. People wonder about a particular mine's depth, its working dates, ownership, and whether it made a profit or not. It is unfortunate that most of these questions cannot be answered because so little written data exists and the few printed accounts are usually in state and government geological reports, somewhat technical and of little general interest to the average reader.

The earliest mining recorded in the Ramapo Mountains, as well as the adjoining Hudson Highlands, was done at Sterling by Cornelius Board after he purchased land in 1736. There may have been—and probably were—earlier attempts in the Pompton area, but nothing has been found. All of the mines except those that were exploratory or test pits have been included in this book and total over seventy-five in number.

The mine holes vary in depth from about ten feet to the tremendous slope of the Forest of Dean mine—roughly 6,000 feet. The latter shaft, now on the U.S. Military Academy Reservation near Fort Montgomery, has been filled in at the entrance, concealing it so effectively that it has become almost impossible to detect.

About 1880 the market for iron ore from this region began to diminish because of the increased use of the Lake Superior ores, mined and transported at lower costs. Many of the mines in the New Jersey–New York area were closed down. By 1900, only a few were still being worked. Gradually the field was narrowed until only the Peters and the Forest of Dean mines were in operation. In 1931, these were closed, ending a period of almost two centuries of continuous mining.

Except for the reconditioning of the Peters mine during World

Interior of the mine. *Harper's New Monthly Magazine,* 1860.

War II by the federal government and unsuccessful attempts to work the same mine after the war by private parties, no further attempts have been made to reactivate any of them. However, tentative plans are being formulated for other uses. In the fall of 1961, for example, consideration was given to using some mines for storage of records; and another for use as large bomb-proof shelters, a suggestion of the civil defense director at Ringwood. There is always a chance, too, that some may be mined again as good usable ore is still to be found in most. However, housing developments rising near many of these old properties threaten that possibility. A word of caution to those who may plan to visit these old mines: it is best not to enter any of the mines described here as lack of repairs has resulted in unsafe timbering which may give way and cause a cave-in. Water in the mine holes can be deep, often several hundred feet. The only safe way of viewing these mines is from the outside on firm ground.

THE MINERS

The life of the typical miner was not easy: hours were long and hard, pay usually small. Charles Green, who was interviewed in 1938 when seventy-six, told about it.[1]

Born in 1862, Green was employed when only nine in the O'Neill mine near Monroe, New York. It was customary then to hire very young boys, a practice which was later abandoned. His first duties were carrying water and drills to the miners, and driving mules that operated a winch used to raise ore from the shafts.

Inspecting the mine. *Harper's New Monthly Magazine*, 1860.

Gallery showing tracks and shoring. *Harper's New Monthly Magazine*, 1860.

There were 300 men and boys working at the mine doing many different tasks. They lived in wooden frame boarding houses, about 40 in all, on the Lake Mombasha road just off the present road between Southfield and Monroe. The houses are no longer in existence.

At twelve, Green became a regular miner swinging a nine-pound hammer until he quit in 1881. A day's pay of 81¢ was considered good: "I kept five persons on my income then better than I could do it today on much more," he recalled. In the Panic of 1873 he received only 45¢ a day, but the cost of living was lower. Eggs were 1¢ each; flour $3 a barrel; and beef and butter each only 12¢ a pound.

The men in the mines wore high leather boots which cost about $3 and lasted about a year of rough, wet wear. It was not uncommon for the miners to stand knee-deep in water when the pumps failed to clear the mine floor. The remainder of the clothing worn by most miners, even in winter, consisted of an undershirt with the sleeves cut out and a pair of overalls. The temperature rarely changed below the frost line and most men washed, and changed their clothing in the engine house before going home.

According to Green the drills, all operated by hammer and brawn, were of three sizes. Each had its special purpose: there was the stoop, or 7½-foot drill, driven into the rock wall a couple of feet off the floor; next came the breast or 5-foot drill, to be sunk midway up the wall; third was the head or 3-foot drill, driven in at head level. When the holes were touched off with explosives, the ore broke off in a scoop-like section.

Green said that by the age of twelve, he used the 7½-foot drill, a two-man operation in which one held the drill while the other hammered it. When it penetrated the rock far enough to stay by

Mule-drawn skip car. *Harper's New Monthly Magazine*, 1860.

A miner. *Harper's New Monthly Magazine,* 1860.

itself, the men alternately struck the drill with their hammers. A day's work was measured by the depth of the drills. A crew usually began at 7:00 A.M. and halted when the hole was ready for the charge, usually just before noon. As soon as the men finished their own drilling they helped the others in their group. Often two crews would race each other to see which team would come out of the mine first.

In pre-dynamite days, blasting crews used a black powder. After the explosion (powder fuses were used before electricity to ignite the charge) muckers came in. They picked up the broken ore, loading it into wheelbarrows and carts to be hauled to the surface. Above ground other men broke up the larger pieces with sledge hammers before filling the heavy carts for shipment to the Greenwood furnace.

A miner found his way around in the mine by the light of a candle carried until he reached his work place. Using a lump of clay, he stuck the light to a rocky ledge in the wall or stood it on the floor in melted wax. The candles glowing along the long black tunnel looked like lightning bugs, Green thought. Shortly before he retired, candles were replaced by kerosene torches which were unpopular not only because of the odor, but they emitted black smoke causing coughing and spitting of black particles. The miners got rid of these torches as fast as they were issued by flattening them with hammers and throwing them away.

Even the rats that infested the mines came to be regarded as friends: no matter how large or how small their candles were at the end of the shift, the miners always left them behind as food for the rats. This repaid the rodents for staying in the mines and warning about potential cave-ins. One day rats in great numbers raced to climb out of Shaft No. 4. Taking their cue, the miners also left. The following night, there was a cave-in in the shaft but no one was trapped thanks to the rats' forewarning. That shaft was never deepened, nor did any of those working it ever wander far from the entrance.

Miners were paid by the month, buying their supplies from the

Pushing an ore car underground. *Harper's New Monthly Magazine*, 1860.

company store at Greenwood furnace. There were seldom any cash transactions as the clerk merely deducted the amount of purchases from a man's name on the payroll. As a result some miners received no pay at the end of the month. Green claimed that the owners were often lenient, however, for if a man overdrew his pay, but needed food or clothing, he invariably received it. Company teamsters took orders and delivered them on specified days. These teamsters also had the job of transporting the ore from the mines to the furnace at Greenwood. Teams of horses could haul 2–3½ tons on every load and make two trips a day, for which the drivers received $1.50 per ton. About 300 loads were carried every day. During Green's time, ore was also hauled to Greenwood from the Bradley, Surebridge, Hogencamp, Mombasha, Bull, and Warwick mines.

The majority of the miners were English or Irish, many of them experienced hands from British mines. In the O'Neill and Mombasha mines both were employed, but the Bull Mine preferred Englishmen and an Irishman did not dare show himself there. At the end of the week most of these miners headed for Monroe Village. "It was hell in Monroe on Saturday nights," Green recalled. A dozen or more free-for-all fights took place in the streets or in the saloons. Several times on Sunday mornings a saloon was found without a door or window left intact, although no one was ever arrested—the nearest jail (at Goshen) was not large enough to hold all the offenders! Some of the miners would customarily dicker with the proprietor on the cost of the damage, usually the next morning, taking up a collection at the mines to pay for the results of their high spirits.

Despite the severe strain of heavy physical labor underground in the mines from 1874–81, Green's life span apparently was not affected greatly, if at all. When he quit the mines at nineteen, he weighed only slightly more than when he started at twelve. Yet, at seventy-six, he was able to recall vividly his mining experience. Of course, there were others whose bodies were unable to withstand the hardship of working in the bad air and otherwise unhealthy atmosphere. Doubtless, the Saturday-night roistering provided needed release from the confining daily work and exhausting drudgery for many of the men, who felt it was well worth the Sunday price.

GREENWOOD GROUP OF MINES

Boston Mine—This mine is located on top of a low ridge about a third of a mile south of Island Pond, near Arden, New York. The mine opening is just a few yards east of an old wood road that runs north to south over the ridge. There is a large open cut in the rock wall through which one can enter to see the water-filled shaft. This mine was last worked in 1880 and, from the size of the dump, does not appear to have been very deep. The ore was sent to the Clove furnace.

Bradley Mine—This mine lies on the Arden Valley Road, four miles east of the railroad station at Arden and a ½-mile northwest of Lake Tiorati. The old mine chamber, with its stalagmites and stalactites, formed by the dripping water, offers an unusual sight. It is readily accessible near the road where a plaque marks the site. The mine supplied ore to both Clove and Greenwood furnaces during the Civil War and continued to operate for some years afterward. During all of this period it was owned and worked extensively by the Parrott brothers. The ore bed was at least 40 feet thick.

Bull Mine—Although some distance north of the other mines in the Greenwood group, the Bull mine is included here because it supplied the Clove and Greenwood furnaces with large amounts of ore. Its location just north of Monroe, New York, and northeast of Oxford, New York, on top of Pedlar's Hill, makes it the most northerly of the mines in Orange County. The ore lies in a shoot which dips to the southeast at an angle of 25 degrees. It is about 100 feet wide, 10 feet thick and is said to have been worked 1,000 feet deep. Over 2,575 tons of ore were raised from this mine in 1880, most of it from an old pillar within 100 feet of the surface.[1] It belonged to the Parrotts and was last worked in 1884. The ore was roasted in heaps before being sent to the furnaces and the iron produced was used principally for foundry hardware. There is a drift at the foot of the hill on the east slope that intersects the

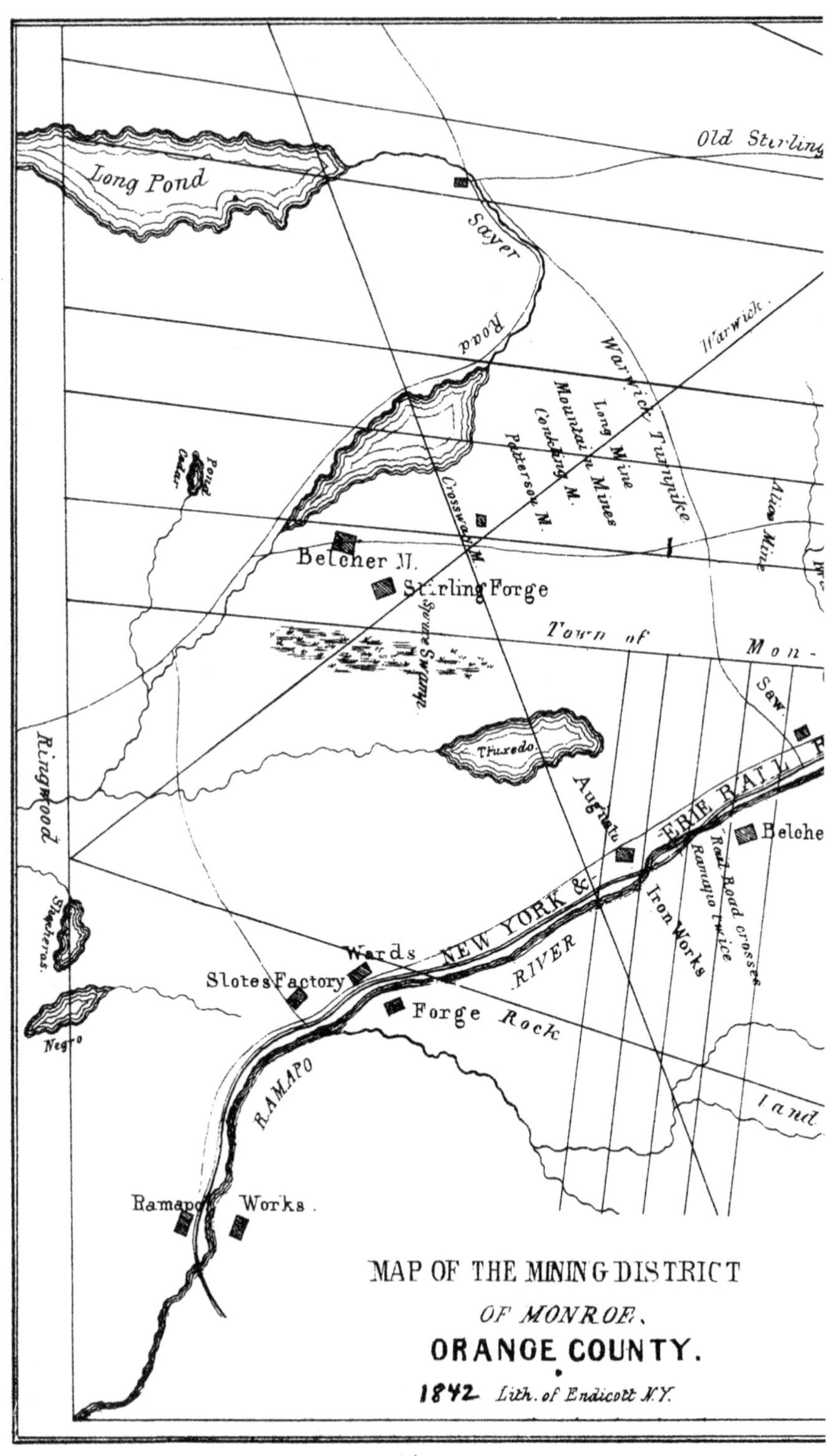

From *Mineralogy of New-York* by Louis C. Bec

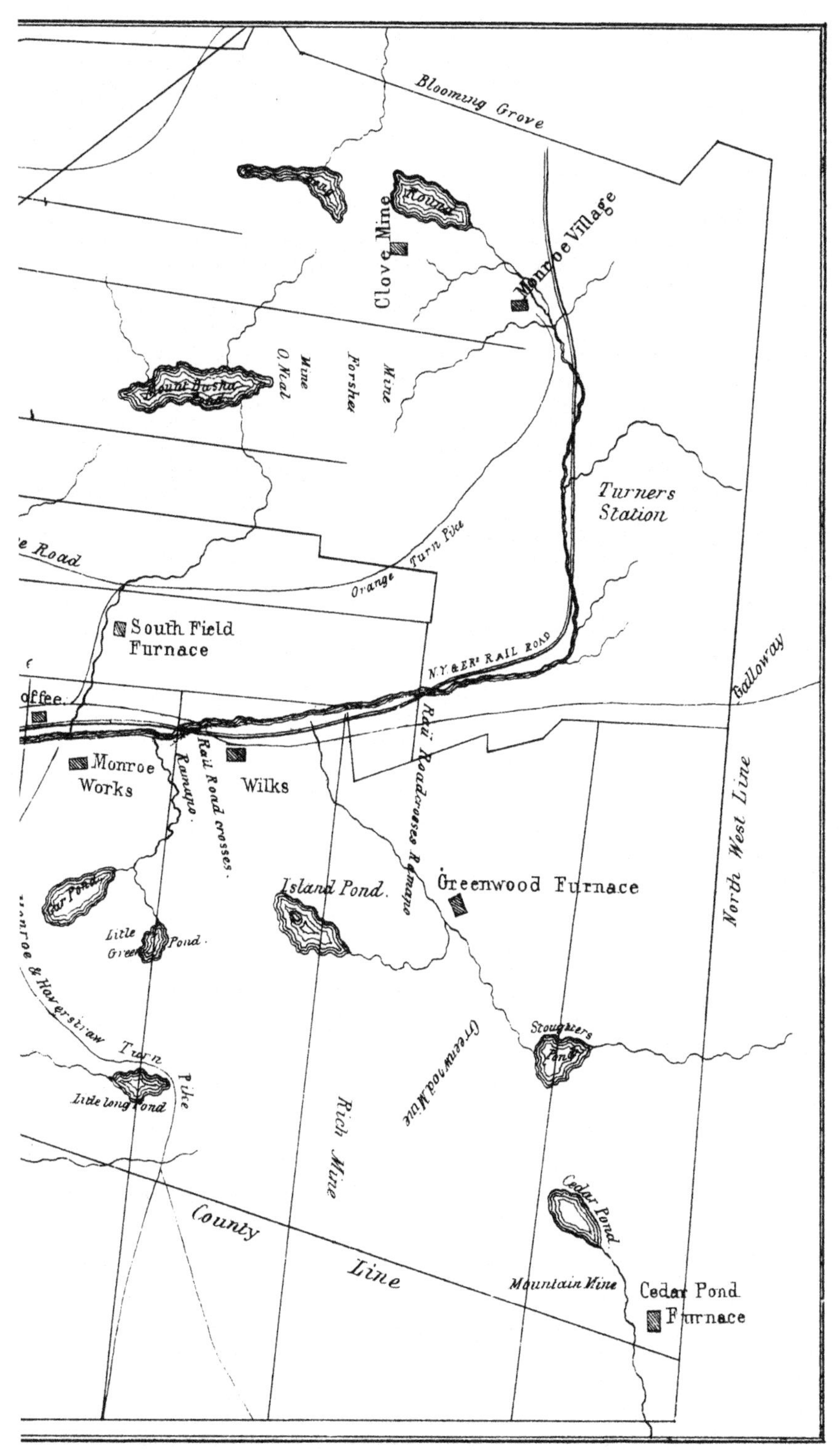

ormer professor at Rutgers College.

Entrance to the Bull Mine. *Photo by Raymond S. Darrenougué.*

shaft, but it caved in and is inaccessible. The lower portion of the mine is water-filled, but the upper part, though dry, should be examined cautiously. The total production of the mine was estimated to be about 52,000 tons by 1881.

Clove or Wilk's Mine—About two miles north of the Forshee mine and 1½-miles south of Monroe is the Clove. It was opened about 1797. By 1838 the workings were over 500 feet in length and evidently extended over a still larger area in all directions. In the nineteenth century, this ore was particularly suited for making "red short" (iron brittle at red heat) or "hot short" iron. Because of its high sulphur content, the ore had to be roasted before being used at the furnace. At one end of the open pit, however, there was a deposit of soft black ore, which could easily be taken up with a shovel and used without roasting. The Parrotts owned the mine from 1839 to 1885, using the ore at their Greenwood and Clove furnaces. In 1880, when the mine was last worked, 15 tons of ore were raised. The workings are now full of water. Besides the large open pit, there are numerous shallow pits, most of which have caved in. The large pit is about 500 feet long and 200 feet wide with a narrow cut about 150 feet long at one end which widens out, leading into an inclined shaft or tunnel of unknown depth, filled with water.

Cunningham Mine—This mine, at Greenwood, is midway between Echo and Cranberry Lakes. In 1880 about 224 tons were raised, but it was never used as extensively as others in the Greenwood group. It was named after its owner, Richard M. Cunningham.

Forshee Mine—This mine can be found 2½-miles west of the Arden railroad station and ½-mile southwest of the O'Neill mine. It was acquired by the Parrotts about 1839 to help supply their Greenwood furnace. The main working was an open pit about 400 feet long, and 50 feet deep.[2] Its ore was highly prized by ironmasters, most of it sufficiently free from pyrites to eliminate preliminary roasting. The mine closed about 1874, but was reopened in January of 1880, and worked until June when it was abandoned. During these months it produced about 2,000 tons.

Garfield Mine—At the south end of Island Pond, this mine was also owned by the Parrotts. One of their smaller and later ventures, it was not worked after 1880. The shaft is now water-filled.

Greenwood (Patterson) Mine—These old workings lie about 2-miles southeast of the Arden station. They were worked as early as 1838. The ore was hard and compact, containing a large portion of iron pyrites, making roasting necessary. The mine had three layers of ore—separated by a few feet of rock—the center vein nine feet thick. This mine furnished ore for both the Clove and Greenwood furnaces during the Civil War and was another owned by the Parrotts. Although not worked from 1870 to 1879, ore was raised in 1880, its last working year.

Hogencamp Mine—This mine of six or more openings was also owned and worked by Robert and Peter Parrott and supplied their furnaces at Greenwood. It was worked from the early days of the Civil War until 1885 when the Clove furnace was finally shut down. The diggings are located about three miles southeast of the Arden station, about a ½-mile southwest of the Pine Swamp mine, on the north side of the Dunning Trail.

The two westerly openings consist of long open narrow cuts extending up the side of the mountain. The one farthest west is roughly 150 feet long and averages almost 5 feet wide. Most of it is filled with water. Another opening, close by on the same vein, is dry but badly caved-in. The next mine hole along the same strike is a dangerous perpendicular shaft with an opening measuring about 10 by 12 feet. Caution should be used when looking into it. A very short distance further to the southeast is a water-filled shaft on the ground level of the trail, just a few feet to the north of the path. An iron pipe of some five inches in diameter, rising above the water in the shaft was apparently used for dewatering the mine. This may have been difficult with the mine located on the north side of a marsh at the foot of a mountain.

In the middle of the trail are some long iron spikes that probably anchored the machinery and boilers which furnished the steam and power for drilling, dewatering, and hoisting. On the south side of the trail one can see a concrete base with more spikes. Charles Jones of Central Valley, New York, remembered seeing a tramway hauling ore up over the top of the mountain when he worked at this mine as a young boy.[3] According to Jones, four gangs of men worked in the mines on the day shift and three at night. Each gang had two or three men, and another six men were used for loading. Two-man gangs could dig out about 8 feet of ore per shift; three-man gangs covered 12 feet in the same time. Pillars were left standing 50 feet apart in the mine to support the roof called the "hanging wall." Four-mule teams were needed to pull 3½ tons of ore, as compared to the 2 tons hauled by a good pair of horses.

To the east of the old water-filled shaft can be seen the foundations of what was once the little mining village surrounding the Hogencamp. These old stone walls are the sole remnants of 20 houses, several barns, a school house, a store and a combination saloon and dance hall. A little stream flows across the trail here running south into the swamp. A few feet east of this stream, about 50 feet north of the path, is a neat circular stone well with the mine dumps nearby. On the east side of these piles of rock can be seen a single rail remaining from a narrow gauge track, where the skip cars ran, bringing the ore from the large open cut lying just a few yards northeast of these dumps. This open cut, about 5 by 50 feet, with sides 30 feet or more deep, seems to have been one of the principal openings.

The Hogencamp workings were the only ones steadily mined in this region from 1870–79. By 1880, the vein had been worked to a depth of 60 feet and a length of 250 feet. That year about 11,500 tons of ore were raised. The mine was never worked after 1885. The vein averaged 12 to 15 feet in width.

O'Neill or Nail Mine—The openings of this mine were located 3-miles southeast of Monroe, New York and about 2½-miles west of Arden, at a spot 4⅕ miles from Route 17 along the old Orange Turnpike at Southfield. It was probably used by James Cunningham to help supply his Greenwood furnace during the War of 1812.[4] In 1823, Gouverneur Kemble, the proprietor of the West Point foundry, leased the mine. On April 1, 1829 he purchased half and owned it all by 1832. Lewis C. Beck visited the mine on June 25, 1838, and remarked that it had been very extensively worked since his visit on April 25, 1837. He also stated that the ore from the "O'Niel" mine was used at Southfield, Woodbury, and Coldspring furnaces.

By 1838 the open pit was already quite sizeable—500 feet long and 150 feet wide. The ore was hard, compact, and favored by ironmasters. It made good "red short" iron. Kemble also owned the Greenwood furnace which he sold to Robert Parrott, in 1839, with the O'Neill mine and other adjoining tracts of land. The ore was mined out around a dike that was left standing like a wall. Great masses of rock have now fallen into the pit.

In the years following the Civil War, 300 men and boys were employed. During this time there were five shafts in all. The longest, No. 5, ran 1,000 feet in length. Just before this mine closed, No. 4 had caved in and No. 5 had reached its practical potential with the methods then used.[5] As a final step, the company planned to strip its pillars which were composed of ore and rock and had been left to support the roof of the mine over the large chambers.

"Shooting the pillars" (removing the good ore) though safeguarded to some extent by placing timbers to support the ceiling load, was dangerous work. Luckily no one was in No. 5 at the time explosives were set off in the tunnel, for the whole shaft caved in making further work impossible.

When the mine was in operation, about 300 loads of ore were sent every day to the furnace at Greenwood. Teamsters and their wagons made two trips daily from the mine to the ironworks, with an average load of 2–3½ tons, for which they received $1.50 a ton.

An examination of the workings in 1921, extending northeast and southwest along the strike of the ore, revealed that they covered a distance of nearly 1,500 feet. In some places the depth was as much as 100 feet and the width varied from 100–200 feet. The main pit, with its two chambers at the eastern end, is about 600 feet long.

Pine Swamp Mine—The group of eight openings to this mine are located a little over a mile southwest from where the Seven Lakes Drive passes the western end of Lake Tiorati. The best way to reach the mine is by taking the Dunning Trail at this point. It is east of Hogencamp Mountain and just south of Fingerboard Mountain. The mine consists of open cuts, pits, shafts and adits, and was probably first worked as early as 1830.[6] Robert and Peter Parrott acquired the mine and transported the ore to their ironworks at Greenwood. The most interesting of the mine holes, the one farthest west, is very easy to find. Its dump rises about 60 feet, close to the north side of the trail. By climbing up and over the dump, one can see the mine straight ahead and just a few yards up the side of the mountain. This sizeable cut is about 25 feet in width, with walls 30 or so feet in height running up each side. The old perpendicular shaft, water-filled at the bottom, is about 10 feet wide and 30 feet in length. In the water several large logs formerly used for shoring (bracing the walls to keep them from falling in) can be seen. The cut and slope runs uphill from the water-filled shaft at a gradual grade, with a hanging wall about 30 feet in height. There is a small opening overhead part way up the slope which lets the sunlight through and helps light the passageway.

In the busier days, this mine like the one at Hogencamp mine, was the center of a small village, with homes, barns, stores, and even a saloon. The foundations of some of these buildings can still be seen. The mine took its name from a swamp lying immediately south. The other openings are mostly water-filled and do not seem to have been worked as thoroughly. All are close to the north side of the Dunning Trail.

One old timer told of a long ramp leading from the main workings to a spot where the ore was dumped before being carted away in wagons. The mine was inactive during the years 1870–79, but it was worked briefly in 1880, before it was abandoned.

Surebridge Mine—This mine is located about two miles east of Arden at a point where the Ramapo-Dunderberg, Arden-Surebridge, and the Appalachian trails pass close to each other. There are two large water-filled shafts. Like the others in this group they were owned and worked by the Parrott brothers, who used them mainly during the Civil War period to supply ore to their two Greenwood furnaces. A total of 458 tons of ore was raised in 1880.

HUDSON HIGHLANDS GROUP OF MINES

Cornell Mine—This mine consists of two openings, both near the summit of Bald Mountain, just to the north of the Ramapo-Dunderberg Trail. Bald Mountain lies just west of Dunderberg Mountain and south of Bear Mountain. It was named for the Cornell family who once resided in the area. Neither opening was worked to any great extent.

Copperas Mine—This old working, opened about 1840 by a New York company, was soon abandoned for lack of usable ore. It was a mile northwest of Fort Montgomery.

Cranberry Mine—Located on the southeast side of Cranberry Hill about two miles west of Bear Mountain. The adit, which was used long after mining had come to a halt for storing dynamite and powder, runs into the side of the hill more than 100 feet. About 100-yards north of the opening is a small stone building probably once used for the same purpose but unused for over 20 years now. One can see where other buildings once stood between the mine and this location. A fair amount of ore was mined here at one time.

Doodletown Mine—This mine hole, probably not much more than an exploratory pit, lies on the eastern side of West Mountain (just to the south of Bear Mountain) where the Suffern-Bear Mountain Trail passes immediately north.

Forest of Dean Mine—This famous old mine was opened at least by 1754 by Vincent Mathews when he bought the property after leasing it in 1753. It was the deepest of all the mines in the Ramapo-Hudson Highlands range, going down about 6,000 feet along the slope. Its ore was very highly regarded and used in markets as far distant as Massachusetts. In Colonial days it was transported to the nearby Forest of Dean furnace and in the post-Revolutionary war period to the Queensboro furnace. The mine lay about 5-miles southwest of West Point and nearly 4-miles west-northwest of Fort Montgomery.

On April 9, 1770, John Griffiths, proprietor of the Forest of Dean furnace, boasted in *The New-York Gazette and the Weekly Mercury,* that this iron ore being used at the Forest of Dean, produced such excellent pig iron, that it was sold for 20 shillings a ton more than the famous Andover iron. With the permanent closing of this furnace in October of 1777, the mine was probably worked on a limited scale until the close of the Revolutionary War and the erection of the Queensboro furnace.[1]

Forest of Dean mine ore was being sent to many localities in the late 1790's. Almost 1,000 tons bringing $7 to $8 per ton were sold yearly for use at ironworks in Rhode Island, Connecticut, Massachusetts, and New York State. In a New York newspaper, *The Spectator,* December 25, 1799, this mine was advertised for sale with the Queensboro ironworks. Interested parties were asked to make contact with E. Luget on the premises or C. Lagarenne in New York.

Eleven years later, in the *New-York Evening Post,* September 14, 1810, the combined Queensboro and Forest of Dean properties were advertised at an auction to be held December 1, 1810. Forest of Dean ore was described in this notice as "so rich as to furnish a ton of bar iron from two and a half tons of ore." The supply was apparently inexhaustible. The ore was estimated to be worth $4 per ton at the mine, $6 at the Fort Montgomery landing and $7 a ton in New York City. It was believed that 2,000 tons could be

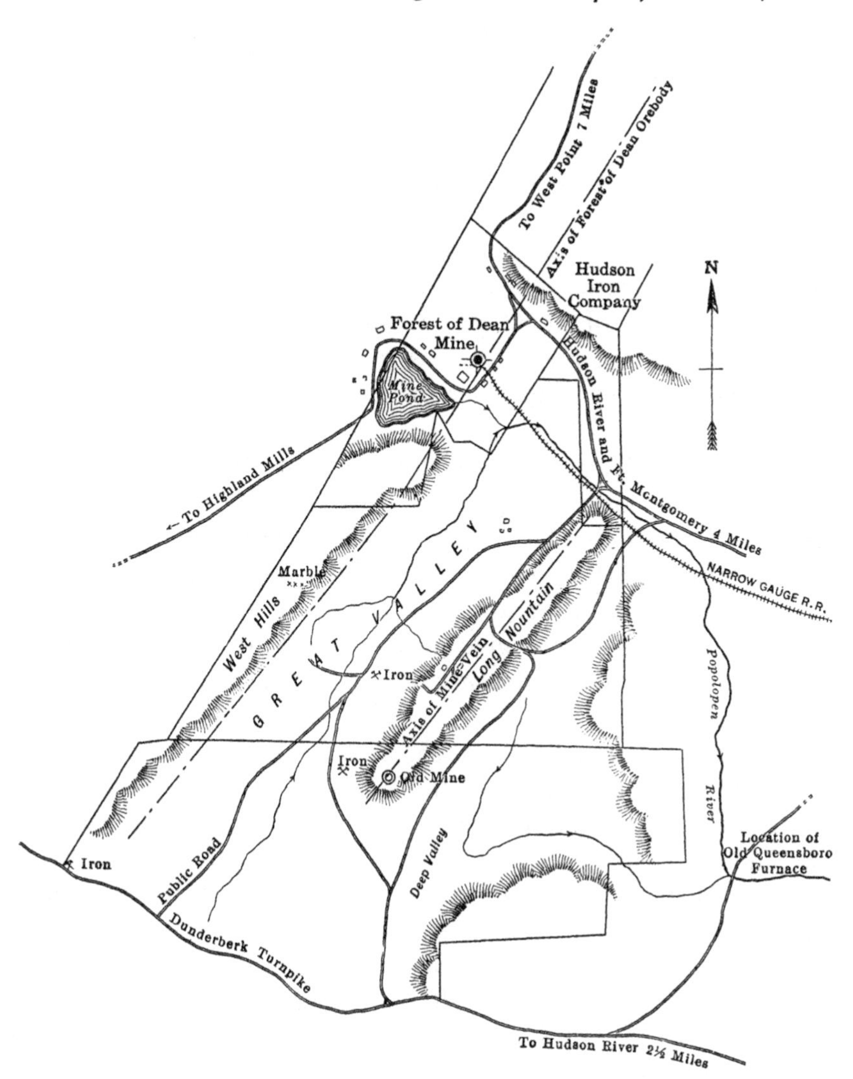

Region surrounding the Forest of Dean Mine.

sold annually. At this time the ore was not used locally to make pig iron, for the Queensboro furnace had been abandoned in 1800.

The mine itself was shut down temporarily in the 1830's, becoming partially filled with water. By 1842 it had been pumped out and was worked under the ownership of George Ferris of New York. A heart-shaped ore vein, 30 to 36 feet in width was opened for a distance of about 500 feet. The depth of the mine then varied from 50 to 70 feet. By 1842, almost 40,000 tons had been taken from the Forest of Dean mine since its discovery more than 80 years previously. Power for dewatering the mine as well as hoisting the ore was supplied by means of a waterwheel located on adjacent Popolopen Creek.

After working the mine for a time, Ferris sold it to the Forest of Dean Iron Company, which was incorporated on November 25, 1864 with capital of $1,500,000 divided into shares of $100 each. The incorporation papers listed as major stockholders: George H. Potts, Frederick A. Potts, Charles C. Alger, Charles Alger, and John Ten Broeck.

About a year later another company, called the Forest of Dean Iron Ore Company was organized, seemingly as a successor. This firm was formed October 12, 1865, with additional capital stock amounting to $400,000. The trustees of this second company were: Edward Beck, John A. Griswold, Albert Town, and J. B. Brinsmade. The company's object was given as the "mining of iron ores, preparing them for market, and transporting, selling and delivering the same."

This company remained in charge of the mine operation for almost 40 years. During this period they mined over 500,000 tons of ore. Part went to the Poughkeepsie Iron Company's furnaces on the Hudson River and some as far as the ironworks in the Lehigh Valley of Pennsylvania. Ore taken from the mine traveled on a tramroad for a distance of three miles; then was transferred into wagons and carted to the dock on the Hudson River at Fort Montgomery. Until the West Shore Railroad came in 1883, the ore was usually transported by boat, both up and down the Hudson from Fort Montgomery.

By 1880 the distance down the slope of the mine had lengthened to 1,200 feet. Water was drained out of the mine by means of an old cornish pump and a vertical shaft nearly 300 feet deep, located about 900 feet to the northeast of the mine opening. In 1880, a total of 23,500 tons of ore was mined. Eight years later, the slope had been extended an additional 150 feet. Two huge waterwheels, 40 feet in diameter, supplied the power for pumping and hoisting. Water was stored overhead in two tanks directly over the wheels. Each tank had a capacity of about 24,000 gallons. One wheel powered the pump, the other the hoist. A single air compressor supplied the power to all drills used by the miners.

The mine seems to have been idle from about 1889 to August of 1894, according to *The Geology of Orange County* by Heinrich Ries published in 1895. H. R. Rogers, New York State mining inspector, visited the mine in 1893 and found it closed. However, when he issued his report in the latter part of 1894, it contained the statement "but work has been resumed since and fifty miners are employed."

Most of the miners here during 1894 and 1895 were probably employed in removing the ore from the supporting pillars. The slope of the mine had reached the boundary line and, the company being unable to purchase the adjoining property, further mining along this shaft was impossible. Their only recourse therefore was to salvage the ore remaining in the pillars.[2] As a result the mine was worked for only about two years and then abandoned. The stripping of the pillars was carried out by the Forest of Dean Iron Company for a new owner, the Port Henry Iron Company.

The mine remained inactive until 1905, when it was taken over by the Hudson Iron Company in October. By then it was so water-filled that approximately a year was required to pump it out entirely, but by the summer of 1906 the drainage was sufficient to permit resumption of mining operations. Among the improvements introduced into the mine was the laying of a track on which cars loaded with ore could be hoisted along the 1,700-foot underground slope. The Hudson Iron Company had also succeeded in buying additional property, and continued working the shaft. A major obstacle was thus removed which had greatly hindered both of its predecessors, the Forest of Dean Iron Ore Company and the Port Henry Iron Company. The vertical depth of the mine at this time was 700 feet.

An excellent description of the Forest of Dean mine by Guy C. Stoltz, a mining engineer, appeared in the May 30, 1908 issue of *The Engineering and Mining Journal.*

> . . . Since 1906 the Hudson Iron Company has been working the mine, paying a royalty; it purposes to advance the workings to its own property, where it is the intention to hoist as before through the Forest of Dean incline shaft.

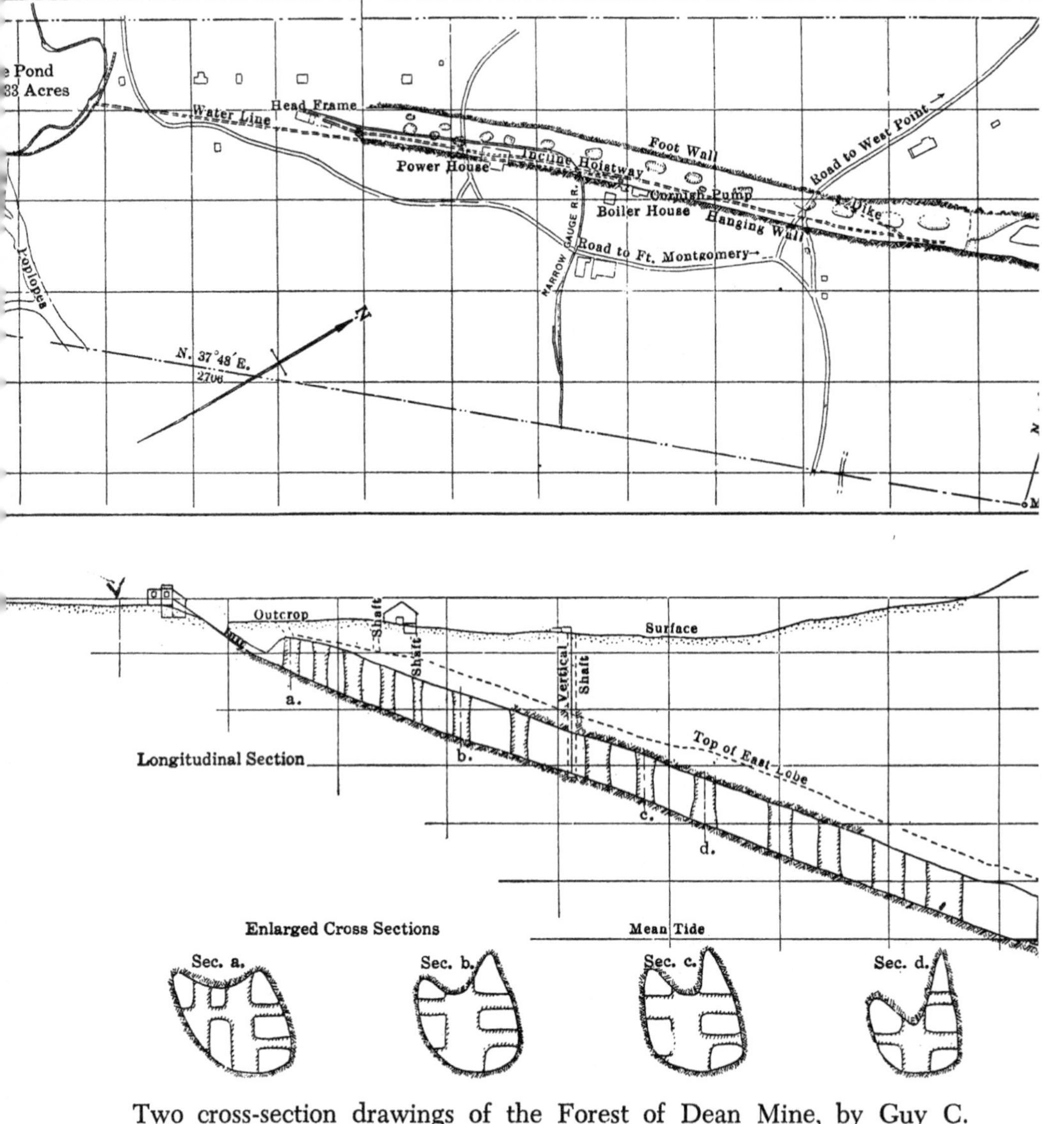

Two cross-section drawings of the Forest of Dean Mine, by Guy C. Stoltz in *The Engineering and Mining Journal,* May 30, 1908.

. . . The deposit is mined by the pillar method recovering about 60 percent of the ore. The ore is hoisted to the surface on an inclined tract at an average angle of 23 deg.; the track follows the footwall along the strike. Timber is used only along portions of the hoistway to protect it from possible falls.

The regular tram cars of 3-ft. gage and 2 tons capacity are hoisted. They are end dumping, and the door is tripped mechanically at the headframe, emptying the ore into a chute, which delivers it, lump and fine, directly to a jaw crusher, capable of receiving a lump 10x20 in.

The crusher is driven by a belt-connected Bacon simple steam engine. After being crushed to about 4 in. the ore is discharged on a 30-in. belt conveyer. Three men pick out the rock and very lean ore, which is thrown into tram cars and taken over a trestle to the dump.

The conveyer empties into an 80-ton storage bin. From the bin the ore is drawn off through three chutes into 10- and 20-ton cars and hauled by a Davenport locomotive over a 3½-mile narrow-gage tramroad and dumped into a storage bin at the terminal of the road. The ore is loaded into the ½-ton buckets (40) of an aerial tramway 6300 feet long, which carries it to a 1000-ton storage bin on the dock at Fort Montgomery. The ore may also be dumped directly into cars or into a boat to be hauled to the company's furnace at Secaucus, N.J.

Both steam and water power are used. The latter is furnished by Popolopen Pond, covering an area of 160 acres, which, controlled by a gate dam, feeds into a creek flowing into the Mine pond covering about 33 acres. A dam provided with a gate valve feeds the water into a 3-ft. pipe line which carries it a distance of 800 feet to the power plant.

. . . I understand the Hudson Iron Company intends to double track the incline and hoist in balance, . . . and to install a generator for lighting the mine and the surface plant by electric current.

By 1921, the mine was being operated by the Fort Montgomery Iron Corporation, successor to the Hudson Iron Company. A total yield of 2,000,000 tons of ore was estimated since 1755. The slope had now been extended to a distance of 4,500 feet with annual production from 1914 to 1918 averaging 88,000 tons. Most of the ore mined by the Fort Montgomery Iron Corporation was sent to Perryville, Pennsylvania, and other Lehigh Valley furnaces. This company continued to work the mine with W. A. Rigby superintending operations until election day in November of 1931, when it was permanently abandoned.

The apparently unlimited resources of this famous mine were emphasized by Rigby's description of an inspection trip he once made of the Forest of Dean mine. "We went down on a slope for about a mile and a quarter, and in that distance the size and shape and dip of the ore did not vary at all. It was the same at the end as at the surface, and there was no reason to believe that it would not go on for many more miles." The final closing of the mine marked the end of a more or less continuous mining period of 175 years—a record scarcely equalled by any other mine in the Ramapos or Hudson Highlands.

Even though activity ceased in November of 1931, water was pumped out of the mine until 1936, and a caretaker lived on the property until at least February, 1938. During a visit made to the mine at this time, personal observations disclosed that some of the families were still living in the area.[3] A long narrow one-story building, formerly a men's dormitory, was still being used as was the little old stone schoolhouse. In December of 1961, an interview with Salvatore Ruscelli of Fort Montgomery, who had worked for 14 years at the mine and lived with his family in the no-longer-used schoolhouse several years after the closing, revealed some interesting first-hand facts about the interior of this old mine.

For six weeks, Ruscelli had been a mucker in the bottom of the mine then, much to his liking, he was assigned work above ground, loading cars with ore. He estimated the underground slope as 1¼ miles in length and the vertical depth about 1,700 feet. There were eight levels in the mine on which the men worked two shifts—one from 7 A.M. to 4 P.M., the other from 5 P.M. to 12:30 A.M. In 1925 a mucker made only 50¢ an hour.

On the visit made to this mine in February of 1938, a tour through the upper part of its vast underground regions gave the author an opportunity for personal observation. The shaftway measured about 10 feet in width and 8 in height; the sidings and overhead supports of timber showed signs of age; in some places railroad rails had been used overhead to support the mine roof. These rails were obviously selected especially for this purpose since some of them were bent under the ponderous weight of the rocks above them. For the first 100 feet down the slope there were wooden steps; after passing the last step, the double track railroad bed was the only path. After a short distance, the shaftway opened into a huge cavern about 50 feet in diameter, in which a small ray of daylight could be seen through the ceiling. This light gave the large chamber an eerie appearance, particularly when flashlights were focused on the rock walls. About every 60 feet, pillars of ore, some as high as 50 or 60 feet could be seen supporting the mine roof in the large caverns.

After a depth of about 1,000 feet, the shaft ahead began to have an unsafe appearance and falling water was heard. Consequently no further penetration was made into the old mine. The decision not to proceed any further into the mine was confirmed as a wise one by the caretaker. He pointed out that a big cave-in had recently occurred at the very spot where the ray of daylight penetrated the ceiling of the first large chamber. The opening to the mine was boarded up soon afterwards as a precaution. At the head of the shaft, hoisting machinery, cables and cars, and a crusher stood, but all trace of the waterwheels had disappeared.

In 1938 the federal government started condemnation proceedings to acquire the property and on March 27, 1942, it officially became part of the West Point Military Academy reservation. This transaction sealed the doom of the old mine, for the opening was filled in. All the buildings were razed and the ground leveled to the degree that one may walk directly over the spot above the old shaft without seeing any trace of it. Completely altered in appearance, the old Forest of Dean mine is inaccessible and lives on only in the memory of those who recall its once commanding and functional role in the iron industry.

Fort Montgomery Mine—This old mine was worked to a limited extent prior to 1840. It was within a few rods of the old landing, on the Hudson River at Fort Montgomery.

Hasenclever Mine—This deposit of iron ore was uncovered by Peter Hasenclever in 1765, who immediately acquired the site, part of a tract of 1,000 acres, lying in Great Mountain Lot No. 3 in the highlands of Orange County, New York. It included a portion of the Cedar Ponds, now merged into Lake Tiorati, which he dammed into a single body of water in preparation for building the furnace nearby. The mine is about ¾-mile southeast of Lake Tiorati.

This 1,000-acre tract was surveyed 126 years later by James D. Christie.[4] At that time nearly 745 acres of it were still known as the "Hasenclever Iron Mining Property." The remaining portion was designated as the "Mary Brewster Lot, Reserved." Mary was the second wife of Samuel Brewster, who apparently worked the mine under a lease arrangement during the Revolution and for some years afterwards. On June 1, 1799, Jonas Brewster purchased the 1,000-acre mine tract from William Denning, Jr., and Thomas Hay, of Haverstraw. Four months later on November 19, Jonas acquired a half-interest in Samuel's adjoining 664-acre tract (including the rest of the Cedar Ponds as well as the brook) where the Cedar Ponds furnace was erected about 1800.

Jonas Brewster died September 4, 1808, and Samuel on November 29, 1821, after which the mining lot property, with the exception of the piece reserved for Mary Brewster, changed hands many times,[5] the mine being worked on and off and apparently unprofitably for the most part. On May 1, 1853, William Knight, who had established a chemical and dye works near the southeast corner of the Cedar Ponds, granted a lease to David H. Mulford: "All that Mine of Iron ore known as the Horse and Clever mine on the Cedar Pond tract of land and all veins of ore connected therewith." Mulford sold the lease to William A. Furnald of New York, who in turn sold it, to begin as of March 1, 1854 and run for 20 years, to the Haverstraw Mining and Iron Company. The Company made the following report in *The Mining Magazine* for February 1854:

> This Company has an estate of about 750 acres situated on the West bank of the Hudson, about five miles from the river and thirty-four miles from New York. The mine was formerly known as the Horsenclever [sic] mine, and was worked some 30 years since, but suspended for want of capital. The mine is supposed to be inexhaustible and the ore according to the representations of the parties concerned, can be mined at 25¢ per ton and delivered on the bank of the river at 75¢ per ton. The ore . . . contains Iron 69.23, Oxygen 21.85 and Sulphur 8.92. About 2,000 tons of this ore has been mined and it has been tested by making it into pig, blooms and boiler iron.

Four months later the Haverstraw Mining and Iron Company came out with a much longer report placing the above-mentioned costs at slightly higher figures. Apparently, this report was an attempt to lure more investors for the company. All of the outstanding shares had not been subscribed for and it was obvious that the project was going to cost more than the original estimates. This report reads in part, "It is estimated upon careful data that the ore can be delivered at the river, all expenses of superintendence, etc., included, at a cost not exceeding two dollars per ton. The directors have the construction of a railroad to connect the mine with the river, in contemplation, which will reduce the transport of the ore to the works and the wharf fifty per cent."

The "works" seem to have been the furnaces being built by the company on a 24¼-acre tract of land along the Cedar Pond Brook (or Florus Falls Creek as it was also called) near the Hudson River at Grassy Point. These furnaces, intended to be placed in operation by June 1, 1855, would be used to manufacture blooms direct from the ore. The report also stated that from 5,000–8,000 tons of ore had already been dug out and was only awaiting the settling of the roads before being transported to the wharf for shipment. Twenty to thirty men were constantly employed in mining and more were to be hired as soon as the demand for ore justified it.

In their appeal for additional investors, the Haverstraw Iron and Mining Company stated that a majority of the stock had already been subscribed. Applications for this ore already mined warranted the expectation that all the ore of ordinary quality, upon being delivered to the river, could be sold at not less than $4 and that richer ore would sell at prices ranging from 25–50 percent more. In the concluding statement of this report a glowing picture was painted of the future prospects.

> The contiguity of the mine to the New York market, superior quality of the ore, the great and growing demand for such ores, its location so near the free navigation of the Hudson, and the economy which coal can be delivered to the works, all give assurance that it is one of the most valuable iron mines in America and it is believed the stock of the Company promises returns affording one of the best opportunities for investment of capital now before the public.

As might be suspected, the Haverstraw Iron and Mining Company failed within a short period, releasing their property back to Knight on December 8, 1856.

The Hasenclever mine had at least four additional owners before it became permanently inactive. In 1857 it became the property of Richard H. Bayard & Son, of Philadelphia. Bayard sold it to Charles W. Galloupe in 1864, who sold it the same year to J. Wiley Edmands, who in turn sold it on March 2, 1875 to A. Lawrence Edmands, an associate of Thomas Edison in iron interests.

In the earlier days the ore from the mine had been used at the Cedar Pond forges and furnace, as well as at some other ironworks further downstream, on Florus Falls Creek.[6] Quite probably the ore was also smelted at the equally old furnace and forges located on Minisceongo Creek in the Haverstraw area. The mine is marked on maps Erskine made of the area in 1777–78, and its original name of "Hasenclever" is still used on maps today to designate the location of the shaft. Since almost two centuries have elapsed since its discovery, that name seems now to be an integral part of the mine, as well as the mountain to the west.

The old mine hole today is about 100 feet deep, cluttered with debris and filled with water to the surface of the ground. It is not an imposing sight, but the surrounding terrain retains items of interest. There are several exploratory pits and trenches, foundations of old barns, homes and other buildings, as well as sizeable dumps of excavated rock. One of the most interesting sights here are the initial excavations and fills made for the road bed of the narrow gauge railroad, lying at the foot of a long, fairly deep trench, next to swampy ground and close to a little stream. This is below the mine dumps. The railroad bed runs for several hundred feet before terminating. Like the company that planned it, this railroad represented a potential that was never fulfilled.

To reach this mine, one follows the Red Cross Trail southwest from Lake Tiorati Brook Road for a distance of a half mile. This trail leaves the highway at a point opposite a small recreation field lying less than a half mile downhill from the outlet of the lake.

Herbert Mine—Located a half mile southwest of the Doodletown opening on West Mountain, this mine lies several hundred yards east of the Suffern-Bear Mountain Trail, and was named after the Herbert family who lived in the area. It was opened before 1859, but never worked to any great extent, being more or less exploratory in nature.

Kingsley Farm Mine—This old opening, mined prior to 1841, is about ¾-mile west of the Hudson River and about 1½-miles southwest of West Point, near the summit of a hill. Tradition has it that lead was found near the mine by the Indians and early settlers.

Kronkite's Mine—This is a very old mine located 4½-miles southwest of West Point and about three miles from Fort Montgomery. There were two veins of ore here separated by a sheet of rock. No less than 800 tons of ore were mined here about 1800 and were said to have made iron of a superior quality. The veins varied in width from a few inches to as much as ten feet in some places, dipping 70 degrees to the west-northwest.

Meeks Mine—Although worked during the 1830's, this mine, situated on the western part of Bear Mountain, was never a major source of ore.

Round Pond Mine—This mine, opened before 1800, is about ¼-mile northwest of Round Pond. From it, a considerable amount of ore was raised, and was said to be very pure in content and excellent for forge work.

About 1834, another opening was made within 100 yards of the northeast section of Round Pond, but only a few tons of ore were mined from this second shaft.

Queensboro Mine—This mine was opened just east of Summer Hill and approximately a mile west-southwest of the ruins of Queensboro Furnace, which it served prior to 1800, when the furnace was operating. After that date the ore for the most part was used at the forges there. When Queensboro ore was mixed with ore from the Forest of Dean mine, the combination made excellent castings. The mine takes its name from the tract of land on which it is located, which was granted to Gabriel and William Ludlow on October 18, 1731. Pre-Revolutionary ore was transported from the Queensboro mine via King's Ferry to the Cortlandt furnace in Westchester County.

Rattlesnake Mine—This old shaft of unknown depth, badly caved-in and more or less water-filled, can be found about ¼-mile southeast of Mine Lake near the location of the Forest of Dean mine.

Smith Mine—Opened in 1828, this mine was between Cro' Nest and Butter Hill. It was located at the foot of the latter, now known as Storm King Mountain. The vein here was from 3–4 feet thick.

Wetherby Mine—This mine hole is about 1½-miles south and slightly to the west of the Forest of Dean mine location. It has an elevation of about 950 feet, on the western slope of the ridge that bounds Deep Hollow on its eastern side. The depth of this old water-filled shaft is not known. The rock structure here strikes north 40° east and dips from 80°–85° southeast according to R. J. Colony.

Zint Mine—This mine lies about two miles northeast of the Forest of Dean mine, three miles southwest of West Point and about a mile southeast of Long Pond. After being abandoned for a long period of time, it was reopened and worked in the summer of 1919. This is apparently the same mine that Professor Colony calls the Tower Mine. He visited there about 1920 and stated that it was 95 feet deep on the slope. He found ore in both walls at the bottom, as well as in the roof of the incline. Directly over the shaft he observed an old drift driven into the rock for about 65 feet, but found no signs of ore in it.

In 1961, Richard Rhinefield, who had worked at this mine in the early 1920's as a driller, told the author that carbide lamps were used and that his pay amounted to $1.50 a day. Fellow employees of his at that time included Sam Vingoe, the dynamiter; Charlie Stevens, a blacksmith; and Charlie Connerly, the teamster who hauled the ore in wagons over the old stage road through West Grove to Benny Havens' dock at Highland Falls. A steam pump was used for dewatering the mine and the ore was hoisted in cars on narrow gauge track. Today this mine is almost water-filled. About a mile to the east, two exploratory pits can still be seen. One is a vertical shaft about 30 feet deep, the other, a quarter of a mile away, is sunk on a gentle slope.

Miscellaneous Mines—William Mather, in his *Geology of the First District,* published in 1842, makes mention of other mines in the Hudson Highlands to the west of the river, but does not give them specific names. One was on the "farm of Isaac Faurot," another, "1½ miles north of Capt. Faurot's"; a third, "1¼ miles north of Capt. Faurot's"; and a fourth, "recently opened on the east side of Bear Hill." Although not strictly a part of the Hudson Highlands group, the Townsend mine in Cornwall, New York, deserves mention here because of the excellent quality of its ore, which was usually powdery or in very small fragments.

RINGWOOD AREA MINES

Blue Mine—Named for the bluish hue of its ore, this mine was opened by Jacob N. Ryerson, prior to 1835. By 1836 sufficient ore had been removed to make the depth 50 feet and the underground length 100 feet. The width of the vein varied from 6 to 15 feet. From 1836, it remained idle for about 15 years until reopened in 1853 by the Trenton Iron Company. By November of 1855 the shaft had sunk to a depth of 130 feet; at this level it extended 60 feet to the northeast and 15 feet to the southwest of the shaft. A tunnel driven in 175 feet from the southeast slope of the hill struck the deposit of ore about 30 feet below the surface. Between 1853 and 1856, no less than 6,000 tons of ore were mined here. The vein varied in thickness then from 10 to 20 feet with its widest point 120 feet from the surface. By 1867 the mine had become 300 feet deep and six years later by further mining it had dropped another 100 feet on the slope, although it was not worked that year. Next to the original shaft, another opening, called the Little Blue mine was made, which by 1868, had been worked to a depth of 100 feet. Its ore was very similar to that of the old Blue mine.

Bush Mine—Located about ¼-mile northeast of the Blue mine and about 1,250 feet south of the Keeler mine, this opening was made about 1854. By 1868 it had been enlarged to 100 feet in length and 70 feet in depth. The vein of black compact ore was 12 feet wide.

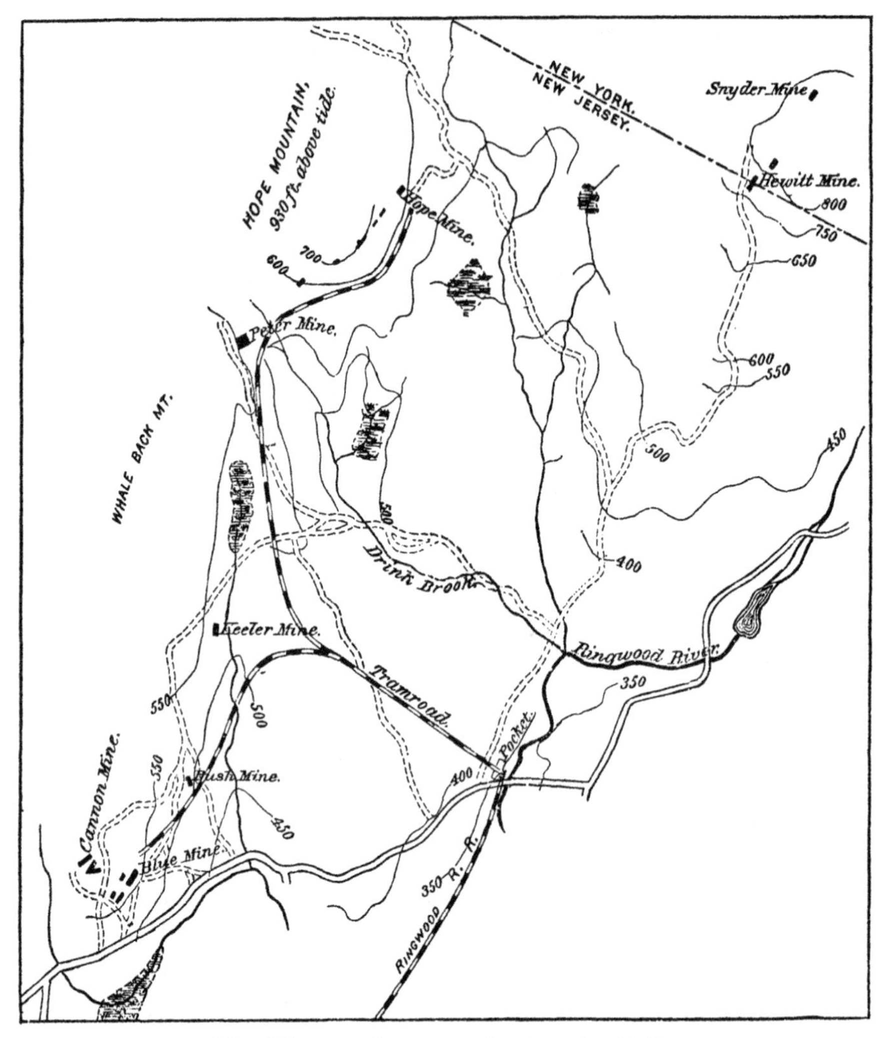

The Ringwood group of mines in 1880.

Cannon Mine—Lying just northwest of the Hard mine, the Cannon openings were, at one time, the most important of the Ringwood mine group. This mine was worked prior to 1763 but was abandoned shortly after the close of the Revolution. It was reopened about 1807, by Martin Ryerson, who acquired the Ringwood property that year. By 1836 the greatest width of ore in the main excavation extended 40 feet, producing a highly brittle or cold short iron. About 1840 the mine was once again abandoned and did not again become active until the Trenton Iron Company reopened it in 1855. It was worked continuously until 1879, and then intermittently for some years. From 1890 on, only the Cannon and Peters mines of the entire Ringwood group were worked.

Within an area of 100 by 125 feet, were four veins of ore, all

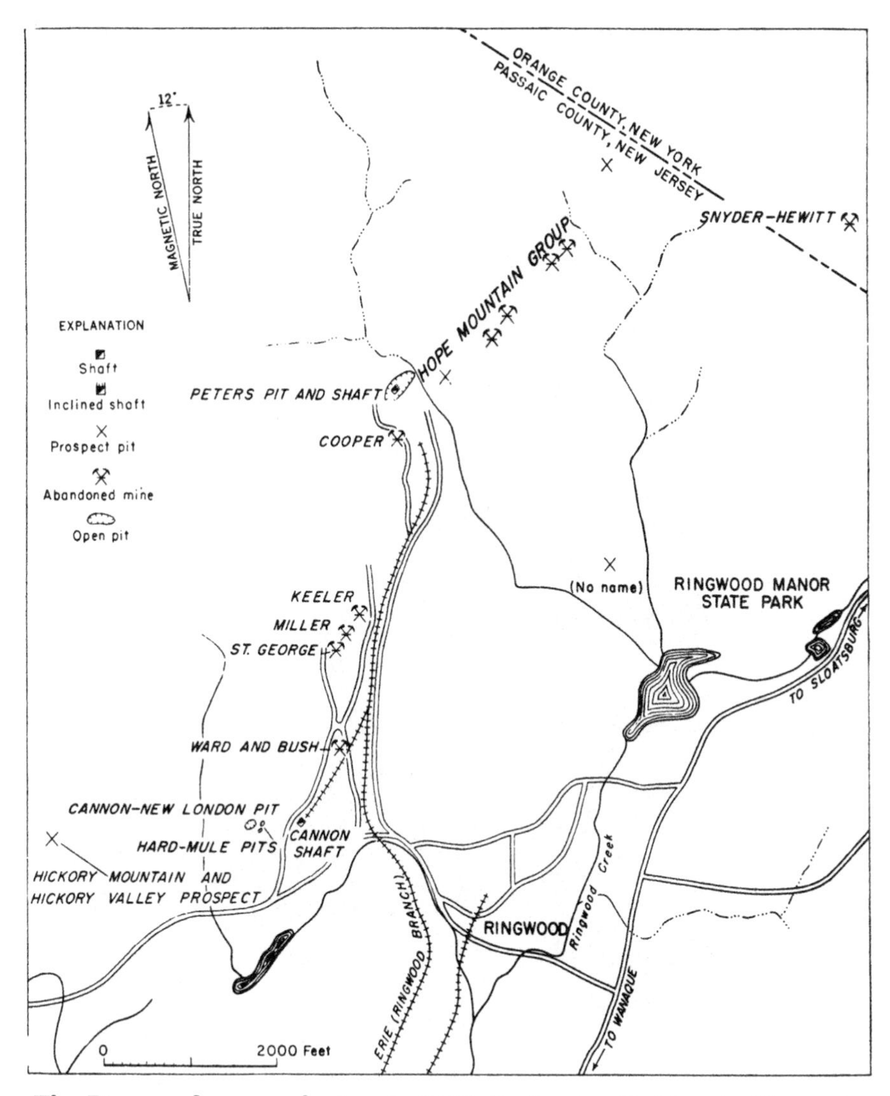

The Ringwood group of mines in 1952. By Preston E. Hotz in *Magnetite Deposits of the Sterling Lake, N.Y.–Ringwood, N.J. Area.*

undoubtedly branches of a single deposit. When the Trenton Iron Company assumed ownership in 1855 the first vein, at the southwest corner of the open works, was 25 feet long, 10 feet thick, and yielded 200 tons of ore. The second vein, in the southeast part, was 50 feet long and 34 feet thick. An opening had been made in this vein many years before 1855 which reached a depth of 25 or 30 feet, but by 1855 it was filled with water and rubbish. The third and fourth veins occupied the northwest part of the

Cannon Mine hoist. *Photo by Raymond S. Darrenougué.*

open works. The extreme northwestern portion of these veins was 60 feet long and 13 feet thick. The remaining portion of these veins was 50 feet long and 20 feet thick. They had originally been opened and worked to a depth of 25 to 30 feet by the old American Company prior to the Revolution.

An 1856 report states that the "ore is both light and dark blue in color." A report appearing 12 years later states that "it is purple when crushed and reddish in color when powdered." There are now four levels in the mine and the total vertical depth is roughly 500 feet. Between the ground surface and first levels, the Cannon mine is a large open pit measuring roughly 150 by 200 feet and 200 feet deep. This opening was abandoned in 1880 and a new one made to the east of the old pit, on a new shoot of ore. This new opening was 60 feet wide. Later, a vertical shaft was sunk to the east which enabled the Cannon, New London, Hard, and Mule shoots to be worked from this single shaft. The mine was inactive from 1884–86, but in 1890 it was in full operation and continued to be active for a number of years.

Some of the interior workings are still accessible, although on a limited scale, and the vastness of the mine's great open pit surprises visitors. The large steel and concrete hoisting tower erected in 1915 which lowered and raised two cars is still standing.[1] The loaded cars were lifted to a point about 20 feet above ground, where the ore was dumped into a chute which in turn emptied it into railroad cars on the track. Silhouetted against the sky the tower, with its two wheels at the top, is a striking landmark.

Cooper Mine—Opened about the time of the Civil War, this mine was located about 500-feet south of the Peters mine. It was worked until 1873. The vein was 10 feet wide, the workings 80 feet long, and in one place the depth reached 30 feet. The opening is no longer visible.

Hard Mine—Opened before the Revolution about 30-yards southwest of the Blue, this shaft was 175 feet deep when it was temporarily abandoned in 1855 by the Trenton Iron Company, who had worked the last 50 feet of it. At the bottom the shaft was worked 50 feet to the southwest and 100 feet to the northeast, producing in two years about 1,500 tons of ore, which was mostly used at the Pompton ironworks. By 1856 the mine was partly filled with water. The ore body dips at an inclination of 60 degrees and has a bluish black color with a bright metallic lustre. The mine was reopened in 1870 and by 1873 it had been worked to a depth of 400 feet.

Hewitt Mine—Lying close to the Snyder mine in New York State and just about a mile to the north of the Peters, this open pit measures about 20 by 60 feet, was last worked in 1880, and is now both flooded and caved-in to a considerable extent. The vein, 6 to 12 feet wide, was worked to a depth of about 25 feet.

Hope Mines—Discovered during July of 1767 on the side of Wales Mountain, later called Hope Mountain, this series of openings lie about 2,500 feet northeast of the Peters mine. By 1768 the vein had been opened in five different places, disclosing nearly a mile-long ore body which was in some cases as much as 14 feet wide. The ore, used at the Long Pond ironworks at that time, was considered to be "a fine tough bar-iron." Among the names given to this series of openings were: the Spanish Hope, Good Hope, Oak, and the Old Hope, the latter being the most northerly located. The workings were abandoned at the close of the Revolution and reopened by Martin Ryerson shortly after he acquired the property in 1807. They were again idle in 1836. At that time, the Old Hope was 60 feet deep and the vein 12 feet wide.

The Oak opening was abandoned at a depth of 30 feet about 1784, although in 1845 it was worked down an additional 30 feet by Jacob N. Ryerson, who later abandoned it. It was still idle in 1856 and filled with water to the mouth of an adit. The ore was compact and black with a metallic lustre. In 1867 the largest of the Hope pits was 30 by 100 feet, of considerable depth and filled with water. All of the Hope workings had been abandoned before this. A little exploratory work was done in the spring of 1880 in an old pit halfway up the hillside but these efforts came to naught.

Today some of the openings can still be seen. There are a few caved-in pits and some timbered shafts filled with water to the surface level on both the side and base of Hope Mountain. Traces of the narrow gauge tracks on which the mine cars ran can still be seen. These tracks ran past the Peters mine and then in a southeasterly direction to Ringwood, where they joined a spur line of the Montclair Railway which ran south to connect with the main line at Boardville. Completed as far as Ringwood by 1878, it provided badly needed transportation. The Peters, Keeler, Bush, Blue, and Cannon mines were also connected by a network of these narrow gauge tracks. Since the terrain provided a down-hill run for the cars from the mines to the junction with the railway spur at Ringwood, the trip down was made by coasting, requiring frequent use of good brakes. Mules were used to pull the empty cars back up to the mines.

Keeler Mine—Opened before 1836, the Keeler mine was just west of the narrow-gauge tracks at a point 2,500 feet south of the Peters mine. The ore from this mine was used for many years by Martin Ryerson at his Bloomingdale forge. The present opening, 20 by 70 feet, and 15 feet deep, is now water-filled. It was idle by 1873 and probably several years before. If this is the same mine that was referred to earlier as the "Caler" (the similarity of the two names would make it seem so), it was opened before the Revolution by the American Company. According to Colonial documents, the company took only a small amount of ore from it. Reopened about 1853, the mine produced 4,000 tons within the next few years, which were smelted at the Pompton furnace, then owned by Horace Gray. From descriptions given as late as 1856, it would seem that the Keeler and "Caler" were one and the same.

Miller Mine—Opened immediately after the Civil War, this mine lies about 375 feet southwest of the Keeler. During 1867 about 1,000 tons of ore were mined here. It became idle in 1873 and was completely abandoned shortly afterward.

Mule Mine—Opened before 1834 by Jacob N. Ryerson, this mine lies about 30-yards south of the Blue mine shaft. Before 1839 it was extensively worked as an open pit, to a depth of 70 feet and a length of 100 yards. The principal vein was, in general, from 7 to 8 feet thick, but widened at some places to 20 feet. It slanted to the north-northeast at an angle of from 40 to 50 degrees beneath a capping of rock. A few feet below the surface at the northeast part of the workings the vein separated into two branches, one continuing in the same direction and the other turning north; the latter branch was worked to a depth of 25 to 30 feet. The northeast branch probably continued across the tunnel leading into the Blue mine. The ore, blue in color and highly magnetic, made good iron, but had a tendency to be "red short" or brittle at red heat.

New Miller Mine—Begun in 1881 and worked until 1884, this mine's vertical depth was 140 feet with a shoot 300 feet in length, 30 feet high, and 12 feet wide. It is now water-filled.

Patterson Mine—Located a mile northeast of Hewitt, New Jersey, in a small ravine, this consists of two flooded shafts and several shallow pits. It was last worked in 1903.

Peters Mine—This mine has been considered for some time to be the most important of the Ringwood group. In all probability, it was first worked by the Ogdens shortly after they acquired the property from Cornelius Board in 1740. When Peter Hasenclever took over the Ringwood ironworks in 1764, however, he found that the mine had not been worked for several years. In 1768 it was dewatered and put into active use again. At the close of the Revolutionary War it was once more abandoned and not reopened until 1807 when Martin Ryerson acquired the estate. In 1837 his son, Jacob Ryerson, cleaned out the old shafts and adits, taking out 1,000 tons of ore. Before this date about 50,000 tons of ore had been mined, most of it during Colonial times. The mine was worked partly in a shaft and partly in an open pit.

While Robert Erskine was managing the Ringwood ironworks before and during the Revolution, he recorded that the two furnaces on the property relied upon this mine for ore. When Jacob Ryerson was working the mine in 1837, the vein was a narrow wall of ore, 6 feet thick, which swelled up into a huge oval mass 50 feet in diameter. This pool of ore had been worked in an open pit to a depth of about 70 feet, drained by an adit cut through the adjoining wall. The ore was said to yield an iron which was brittle at red heat, or "red short."

Thirty years later, the mine was not in operation but the opening showed very extensive former workings, having been expanded to 50 by 130 feet. Its depth was hard to gauge, the sides having fallen in, but the adit from the foot of the hill was visible. It was still idle in 1873, but was working in 1880, and in 1884 a new slope called the New Peters was sunk northeast of the old opening. By 1886 only the Peters mine of the Ringwood group was being worked. At this time an immense quantity of ore was coming from the mine and being placed in piles but little of it found its way during the year to the furnaces of the owners at Durham, Pennsylvania, via the New York and Greenwood Lake Railway.

In 1890, only the Cannon and the Peters mines were in operation, the latter now extending 600 feet down the slope. An observation was made at this time that "the Peters Mine is worked out and will be abandoned as soon as the pillars can be wholly or partially recovered. To recover these pillars the mine is being uncovered, the depth to which the cap rock is now removed, November 28, 1890, is about ninety feet." This prediction turned

Miners at the Peters Mine, circa 1925.

Peters Mine and processing plant, 1940. Buildings replaced by the government during World War II.

Peters Mine, 1940, before being dewatered by the government. Note the single tracks.

out to be far from true. By 1896 the mine shaft had been pushed 800 feet down and four years later it produced 18,755 tons of ore—an impressive volume. The shoots of ore at that time were from 15 to 60 feet wide. On the ninth level an exploratory tunnel was dug leading to Hope Mountain and the ore zone extended 2,500 feet to the southeastern slope of the mountain.

In these days the ore was taken in wooden railroad cars from the Peters mine to the New York and Greenwood Lake Railway spur. The ore-filled cars would go joined, six at once, a hand brake for every two cars. A young Negro was assigned to sit on top of each car with a brake. A dinky engine pushed the cars up to the top of a hump in the track, a short distance from the mine, then the cars would go by gravity, the three boys guiding them with the brake on an exciting ride all the way down to the junction with the railroad. The engine came down later in the day and hauled the string of empty cars back to the mine. In the earlier days, horses and mules pulled the empty cars back.

The mine was worked on and off until June 15, 1931, when it became the last in the Ramapos to be closed. However, it came to life temporarily during World War II, when the federal government took it over in 1942, and had the Allan Wood Steel Company put it in a standby condition. The old buildings connected with the mine were taken down at that time and new modern ones built. Dewatering of the mine started in September, 1942, taking seven months to complete. The entrance to the shaft was rebuilt by putting down a concrete shaftway for the first 200 yards.

On September 8, 1944, the government superintendent at the mine revealed that there were 17 levels in the mine, that the shaft went down 2,400 feet on an incline, and that the bottom of the mine was then 1,800 feet below ground.[2] Originally there was only one double track. The shaft had to be widened to accommodate an additional one, so that an empty car could go down one track for ore while a loaded car was coming up the other. During this renovation, an old wooden water pipe was uncovered at the bottom of the mine which apparently was installed in Colonial days. A stairway was also found that went to the bottom of the shaft. The mine was actually operating for a very brief period during World War II, but all work was stopped on orders from Washington because of transportation difficulties. A number

Peters Mine ore processing plant built during World War II. *Photo by James Skyrm.*

of new features had already been added to the mine plant including a new mill, a magnetic concentrating building, and an electric hoist.

The rebuilding and reconditioning work is said to have cost the government $3,923,000 yet very little ore was mined for this enormous sum. Toward the close of the war, the government brought up some additional ore for testing purposes only. The Ringwood Company supposedly received $300,000 for the mines in 1942. Three years later, they were declared unessential and transferred to the War Assets Administration. On February 6, 1945, the Borough of Ringwood criticised the closing of these mines after two years of work and the expenditure of tremendous funds.

In April of 1947, the Ringwood mines were sold to H. P. Moran, a New York engineer, for a sum estimated at $1,297,000. A report circulated at this time foresaw the reopening of the mines on the first of June. It was planned to ship the ore mined by the new firm, known as the Ringwood Mines Company, Incorporated, to the Bethlehem Steel Company's furnaces in Pennsylvania. Unfortunately these plans were never carried out and again the mines closed down,[3] and reverted to the government in March, 1948. In July new bids for the property were received by the government with the mines being sold in December to the Petroleum Export-Import Corporation of New York. This company, according to reports, planned ultimately to mine as much as 400 tons of ore a year at Ringwood, but once again the property reverted to Washington.

Although a top offer of $400,000 was made for the Ringwood property in 1949, this bid was turned down. During July, 1951, the Ringwood Iron Mines, Incorporated, whose principals included Colonel Lewis Sanders,[4] came into possession of the mines, buying them from the government for a figure quoted as $1,500,000. According to reports, they intended to manufacture powdered and pelletized iron. On September 3, 1954, this company closed operations due to the condition of the iron market. A statement to employees read in part,

> For the past four months we have continued operations, meeting all our weekly expenses without any revenue coming in. The uncertainty

New double track installed during World War II. *Photo by James Skyrm.*

Four views of the bottom of the Peters Mine, 2400 feet down. *Photos by Raymond S. Darrenougué.*

The seventeenth, and lowest, level. See facing page.

> of the future market necessitates our closing down the plant. All shifts will complete their work week on Friday, September 3rd. The Company regrets deeply having to take this drastic step but feels it necessary in order to protect its cash position so as to enable it to meet future demands when the time comes. We wish to express our thanks to all those who went through this trying period with us and we certainly hope to see all of you in the near future back at Ringwood. This is merely a temporary situation and as soon as the market shows signs of returning to normalcy we will start operations again. We hope that when this situation is reached we will have devised a product that will be saleable at any time and in any kind of a market. Thank you again for your co-operation. Ringwood Iron Mines, Incorporated, Richard I. Goodkind.[5]

In July of 1955 the government obtained a judgment against Ringwood Iron Mines, Incorporated, and seized the properties. An auction was held on the property in December of 1955, but there were no bidders. The following year, the Borough of Ringwood foreclosed pointing out that over $100,000 in back taxes was owed them from the property's former owners. At one time the mine was considered for use as a storage warehouse, the ninth level consisting of a cavern that could provide as much as 14,000 square feet, with 1,000 feet of solid protection overhead. On December 4, 1956, the Government was the only bidder, buying the property at a price of $1,685,367.43, only to sell it again at auction on February 28, 1958. The Pittsburgh and Pacific Company took title on July 16, 1958.

The following three years the new owner kept the mine on a standby basis, with the pumps going and a watchman on duty. By the fall of 1961, however, both the dewatering and the watchman's job had been discontinued. All the machinery and items connected with a mining operation of this magnitude, were put up for sale, tagged with various lot numbers for potential buyers. Even the narrow gauge rail tracks that ran from the mine south to connect with the spur of the old New York and Greenwood Lake Division of the Erie Railroad, over which the ore was carried for years, were taken up and made ready for buyers, who would find it easy to melt the rails, making them over into new trackage of a different size or gauge.

The company is also planning, according to latest reports, to take down the trestle supporting the double track running from the hoisting building into the mine itself. Most of the wood in the trestle is so rotten that it is not resaleable, nor are most of the machinery and buildings put up by the government during the war and now obsolete. In an interview with the author, November 18, 1961, Russell Whitney, an agent of the company, said that the operations of processing the ore in the buildings were located so far apart from each other that they increased the costs of production. He considered all the buildings to be worthless, stating that the only reason for buying the property had been the potential value of the iron ore still in the mine. Evidently the present plans of the Pittsburgh and Pacific Company envision building a modern new plant and reconditioning the mines, with the use of modern motor trucks for carrying the ore to markets; but this is all subject to the price of iron rising to a level where such a large outlay of capital would be justified.[6]

St. George Mine—Opened prior to the Civil War, this mine lies 250-feet southwest of the Miller mine. Idle from 1863 to 1873, it was probably abandoned permanently in the latter year. The vein of ore was 30 feet wide. The pit is 35 by 60 feet, 30 feet deep, and now flooded.

Snyder Mine—This pit is a mile north of the Peters mine, just over the New Jersey border in New York State. It was opened prior to 1835 and had been worked by that date to a length of 100 feet and a depth of about 25 feet. The ore was sulphurous and produced cold short iron of comparatively low value. It is now badly caved-in, about 150 feet in length, 25 feet deep, and filled with water. As far back as 1889, it was said that the mine had not been worked for years.

Wood or New Mine—This mine, opened in 1854 by the Trenton Iron Company, is 400-yards northeast of the Blue mine. The vein, including five feet of rock separating the deposit, measured about 12½ feet in width. In little more than a year the Trenton Iron Company took out over 300 tons of ore.

STERLING GROUP OF MINES

Antone Mine—Discovered and opened by John Antone about 1800, this opening actually is a continuation of the Crossway mine described later in this section. In 1838 the Antone mine yielded ore that had about 50 per cent iron content. It was quite malleable with good casting properties, although the cost of mining was as much as $1 per ton. Approximately 50,000 tons of ore were taken out of this mine. The ore body averaged from 5 to 8 feet in thickness.

Augusta Mine—Lying 1,400-feet southwest of the Cook mine, these workings today are caved-in, flooded and inaccessible. Only three flooded shafts and two water-filled open cuts can be seen at the site today. Two shoots were worked here. One, 400 feet long, the other 250 feet. At its widest point the vein measured 9 feet. Both shoots were worked to depths of 175 feet. At a point 200 feet below the surface under the first shoot is a 400-foot drift used for exploratory purposes. From 1859–65, the Southfield furnace used ore exclusively from the Augusta mine. "Augusta Ore" can be found on the pages of the Southfield Blast Account Ledger. In 1880, the mine yielded 560 tons of ore before being temporarily abandoned. About 1894 the Augusta mine was reopened, but it is not known when it was last worked many years ago.[1]

Bering Mine—Located 1½-miles southeast of the Morehead mine in Orange County the Bering mine is close to the Rockland County borderline. It is badly caved-in and filled with water. There is an open cut about 300 feet in length and nearby a funnel-shaped vertical shaft of unknown depth, which narrows down quickly from 30 feet to 10 feet in width. The vein was 15 feet thick at its widest point and was worked through a series of cuts. In 1880 it produced 2,688 tons of ore. Ore mined here was used at the Southfield furnace from August–November 1881, when the mine was abandoned. The old workings lie 500 yards north of Eagle Valley Road, where the Sterling Mountain Railway spur to the mine once crossed.

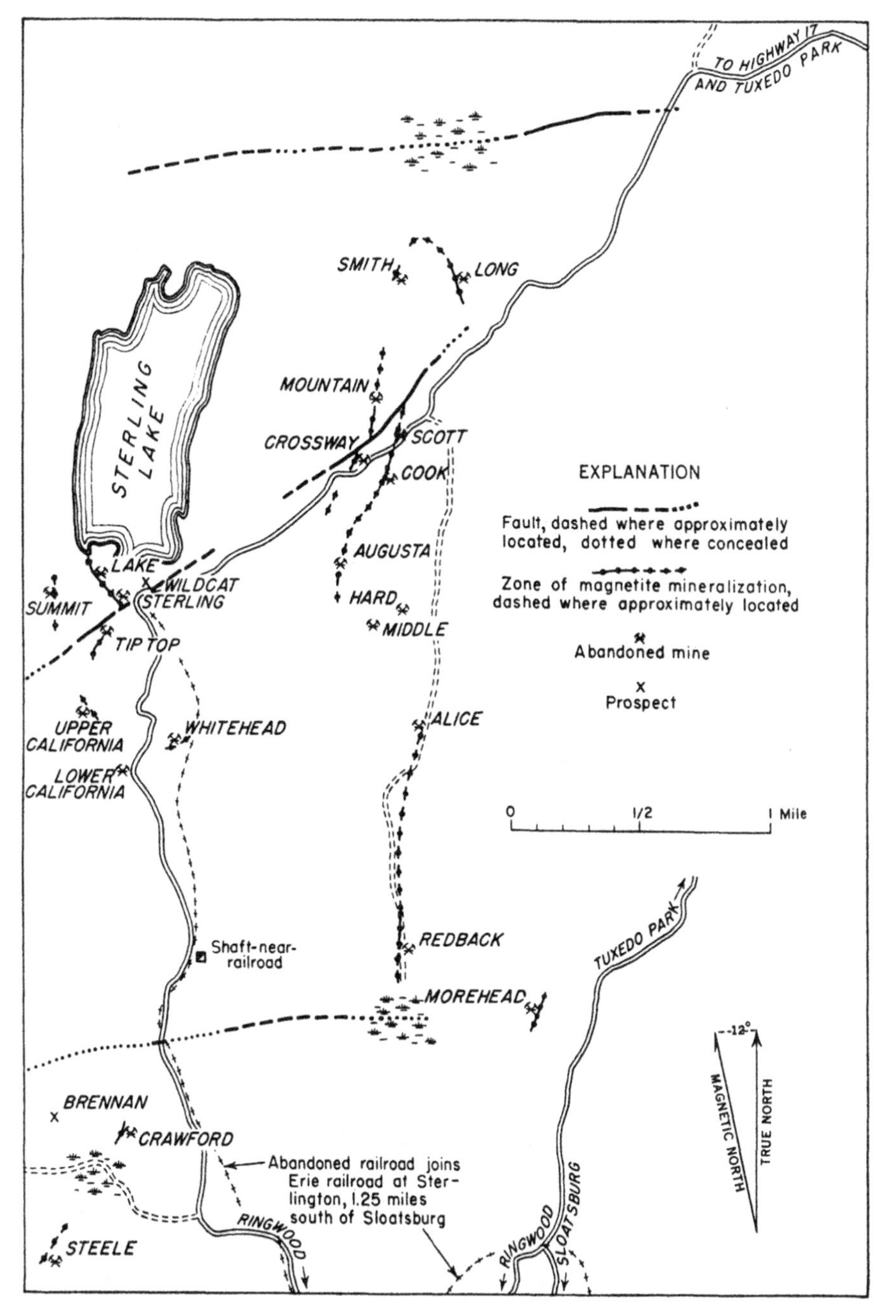

Sterling group of mines, 1952. By Preston E. Hotz in *Magnetite Deposits of the Sterling Lake, N.Y.–Ringwood, N.J. Area.*

Brennan Mine—Comprising one flooded shaft and four shallow pits, these workings can be found about 1,600-feet west of the Crawford mine. The shaft is 20 feet deep and the ore vein about 10 feet in width. There is an abrupt termination of the ore body at that point due to a fault in the rock. No doubt this is the reason why the mine was never worked extensively.

California Mines—Two mines known by this name were worked as early as 1852. One is called the Upper California, the other, the Lower California. The former lies about 1,500 feet south of the Tip Top mine and is at the base of a hill. The shaft sloped to a depth of 350 feet before operations were halted. It is now flooded and inaccessible. The Lower California opening is about 800 feet southeast of the Upper counterpart in a swamp just west of the road running from Sterling Lake to Sterling Pond. In 1921 it was completely under water. This old mine had been reopened by the Sterling Iron and Railway Company some years before, but proved to be so extremely wet due to its location, that further operation was impossible.

Although it is situated in the swamp, the timbered vertical shaft of this mine was visible in the fall of 1965. The opening (6′ x 8′) was filled with water to within inches of the surface. Tailings from the mine were also visible nearby in the swamp.

Cook Mine—This was probably named after Jay Cooke, one of the controlling partners of the Sterling Iron and Railway Company, but the "e" appears to have been quickly dropped for one reason or another. The mine was opened about 1866, the same year its ore was first used at the Southfield Furnace. By 1889 it had been extensively worked to a vertical depth of 250 feet. Due to its location on fairly low ground, the Cook mine had more difficulty with water than some other mines in the same area. In 1921 the work of dewatering the mine was in progress. The Ramapo Ore Company, which had leased the Sterling property in 1918, decided to work both the Cook mine and the Scott mine, there being an underground connection. For this purpose a new 600 foot vertical shaft was put down 100 feet west of the Scott ore body. Despite all the preparation and expense, neither mine was used after 1923.

Entrance to the Upper California Mine, now caved-in. *Photo by Raymond S. Darrenougué, 1952.*

Just inside the entrance. *Photo by Charles F. Marschalek, 1952.*

Crawford or Belcher Mine—This ore vein was discovered by Jacob Belcher in 1792. The mine was known by his name until shortly after the Civil War when it was named for one of the partners, David Crawford, Jr., of the Sterling Iron and Railway Company. The cost of mining here in 1838 amounted to $1 per ton. The ore averaged 48 per cent iron content, was cold short and well adapted for making bar iron by the blooming process. At the time of the name change, it had been worked about 115 feet along the vein. In 1880, the yield was 2,240 tons of ore averaging 57.66 per cent iron.

The mine lies about ½-mile northwest of Sterling Pond and 1,600-feet east of the Brennan mine. The ore was 35 feet thick at the center, 125 feet high in the vein along the strike and dipped to 80 degrees to the southeast. From 1880–87, the ore from this mine was used at the Southfield furnace. The mine was abandoned about 1890, when the newer Sterling furnace was preparing to go out of blast for good. Today one can see that the mine had been extensively worked. There is a large open cut, 375 feet long, 15–20 feet wide, and 40 feet deep. A few smaller pits can also be seen in the vicinity.

Crossway or Causeway Mine—Discovered by John Ball in 1793, this mine was worked until 1809 when it was temporarily abandoned due to wetness. During this 16-year period about 28,000 tons of ore were taken from it. The mine can be found 800 feet northwest of the Cook mine. It yielded ore about 50 per cent iron content which was suitable for casting and malleable. When abandoned in 1809, the ore bed was 14 feet thick. The mine at that time was 65 feet deep and about 150 yards long. The ore vein and walls of the cut were nearly vertical. It was re-opened and worked in 1857 for a brief period. The ore was moderately red short and very similar to the type found in the Patterson mine. Today, an open, flooded cut 10 to 15 feet wide and over 100 yards long can be seen running north-northeast from the Sterling Road. The depth of this cut is nearly 100 feet and approximately 78,000 tons of ore have been removed since it was first opened.

Lake Mine—The date of the opening of this important mine is not known. Although it is said to be a very old one, several detailed reports made by geologists of the region, from 1836 to 1843, make no mention of the name. It lies just north of the Sterling mine, a few yards from the southwest shore of Sterling Lake. By 1888 the mine had been worked only to a depth of 130 feet. At this time the distance between the nearest level of the Lake mine and that of the Sterling mine was only 75 feet. The ore was raised in skip cars up out of the shaft and dumped into the company's railway cars where it was shipped without further handling to its destination. Until the last of the furnaces closed down on the property in 1891, some of the ore was used there, the rest sent to the furnaces in Pennsylvania.

The Lake mine was a very large producer, the only one of the Sterling group to be worked continuously from 1896 through 1917. Total output amounted to 1,254,283 tons of magnetite ore. In a report made by Thomas Pearce, the mine foreman, for the period of January 1 to 6, 1912, the following figures reveal the amount of manpower required for the mine's operation: 20 men drilling by power; 19 men loading ore; 44 men mucking; four miners' bosses; two compressor engineers; three hoisting engineers; two electric locomotive engineers; three pump and bell men; four timber men; five men for general work other than mining; one carpenter; four blacksmiths; five men for positions of timekeeper, shaftman and teamsters. At this time the Lake mine worked on three shifts and produced and shipped 806 tons of ore during the six day period. Lewis Sanders, a mining engineer, had been hired to work on and direct the "Lake Mine Improvement" project which was also going on at this particular time. Forty years later he was part owner of the Ringwood group of mines and operated the Peters mine for a brief period.

The majority of miners who worked in the Lake mine during these years were Italian. One section of the area where they lived with their families was called "Little Italy" and many of the signs posted here were entirely in Italian.

The Ramapo Ore Company leased the property and worked the Lake mine from 1918 to 1921 when they abandoned it, despite the fact that they had completely dewatered it and installed new equipment, as market conditions did not justify further operation. The ore body at this time was 500 feet in width. Wages at the mine declined until the average miner's pay dropped from $8

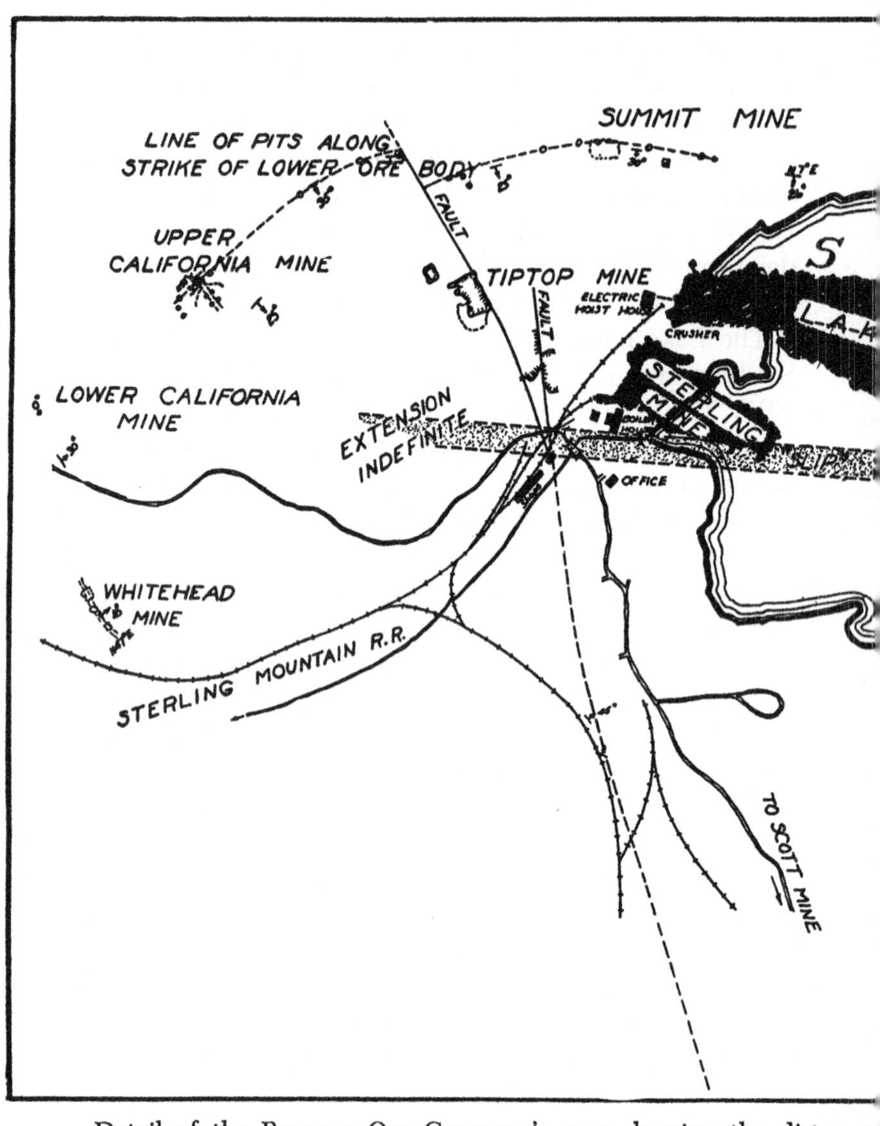

Detail of the Ramapo Ore Company's map showing the distance

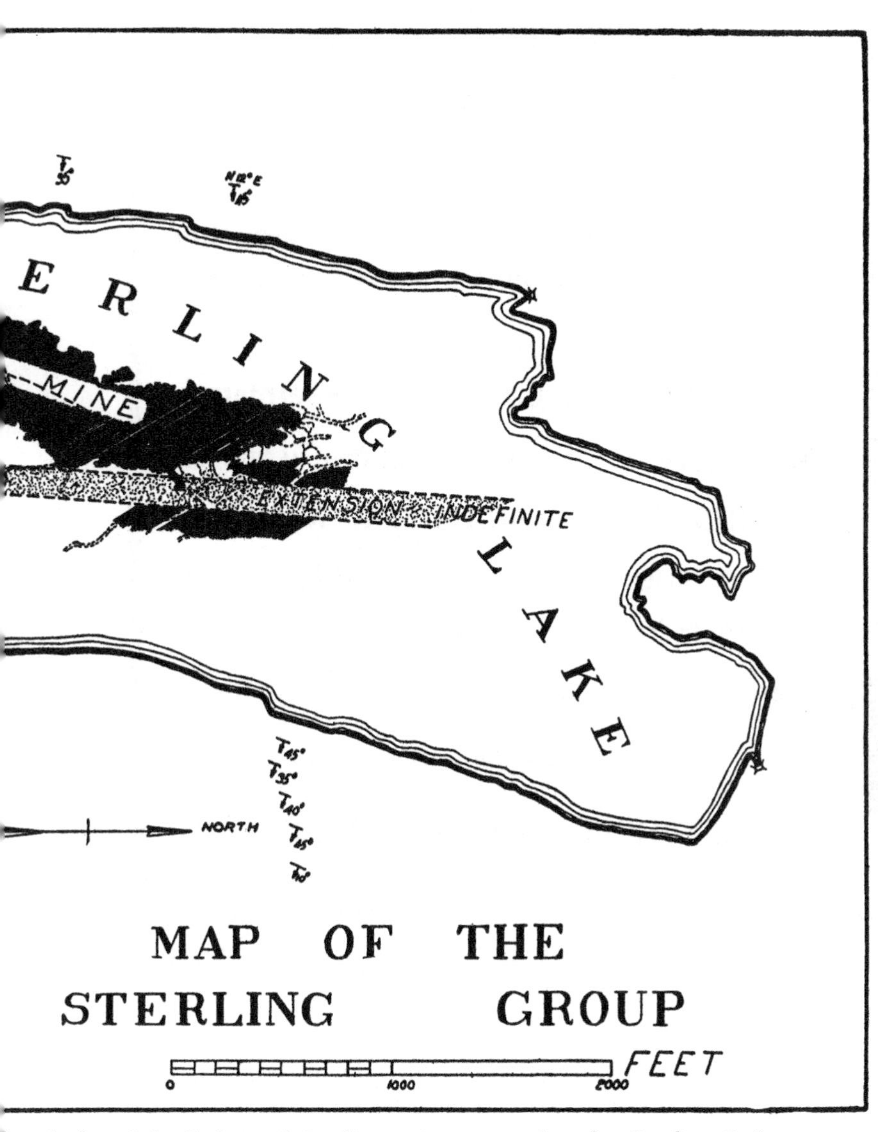

he shafts of the Lake and Sterling mines extend under Sterling Lake.

AVVISO.

SEGNALI DI ALZATA da essere osservati in tutte le miniere della Compagnia Sterling Iron and Railway Company.

Una suonata di Campanello—Fermate.

Due suonate di Campanello—Giù adagio.

Tre suonate di Campanella—Alzate via.

Quatre suonate di Campanella—Alzate adagio-uomini a bordo.

E' proibito montare sui skips o carri da miniera.

Per ordine della Compagnia.

THE STERLING IRON & RAILWAY COMPANY.

Cloth poster in Italian, used at the Sterling mines. It stipulates precautionary measures to be taken.

to $3 per day.[2] As many as 400 men were dismissed, it is said, in one day at Sterling.

Today the shaft of the Lake mine is water-filled. It inclines down at an angle varying from 12 to 25 degrees, for a distance of 3,800 feet to the bottom of the mine, well out under the lake. As a result the lowest level of the mine is about 1,000 feet below the lake. Drifts were driven from the slope at various levels, some going as much as 1,000 feet in length. A precautionary measure,

Boarded-up entrance to the Lake Mine. *Photo by Chester Steitz, circa 1935.*

during the working days of the mine, was to have a check made of the distance between the roof of the mine and the bottom of the lake to be certain that there was an ample safety zone between the two. This check was, as a rule, made yearly by an outside expert hired for this specific purpose.[3]

In its last period of operation the mine was equipped with an electric hoist, a magnetic concentrating mill with a capacity of 50 tons an hour, and dryers. Most of these buildings are in ruins today.

A new road, recently laid down in this area from the southern end of Sterling Lake along its western shore, passes by the remains of a dryer and the concentrating mill. Two other buildings connected with the mine, including a tool and drill house built entirely of small pieces of black rock, were taken down to make room for the road. This building had been used to store samples of the cores (cylinder-shaped pieces of rock), drilled and taken from the various Sterling mines. The new road also passes within a few feet of the opening of the Lake mine. The narrow-gauge tracks that once carried the ore-filled cars can be seen under the water, in the mouth of the mine, extending downwards. There seems to be much ore left in this working, but the present owners have other plans for the property than mining.

Long Mine—This mine, discovered in 1761 by David Jones, has never been "bottomed." It lies about 4,000-feet northeast of the Scott-Cook mine's vertical shaft hoisting tower, and can be traced over a mile in length. An open cut 700 feet long and several smaller shallow cuts, making up the workings, can still be seen today. In 1837, Lewis C. Beck reported that the Long mine had been very extensively worked, but owing to injudicious management, the old workings were then closed. He said the ore was highly esteemed because it furnished pig iron which was easily malleable. An old Southfield furnace ledger indicates that 1,072 tons of ore from the Long mine were used there between October 20, 1834, and December 29, 1836, practically to the exclusion of ore from any other mine.[4]

Two years later, Dr. William Horton, an assistant geologist of the first district of New York State, who lived in Craigville, Orange County, New York, wrote:

> Long Mine is about five miles southwest of Southfield Furnace, and two and a half miles west of the saw works. This mine has been known since long before the Revolution, and is the only one in the County at which anything like systematic mining has been attempted. Only one layer of the ore has been worked, which is six feet thick. The ore is strongly magnetic and said to be as good as any in the Highlands. It requires roasting and gives a red short iron. This mine has been worked three hundred yards in length and about forty in depth. The early working has caved in; but the fine pillars of ore sustain the hanging wall where the work has been recently prosecuted. Vast quantities of ore have been taken from this mine without any appearance of its failing. It belongs to the Messrs. Townsend.[5]

Peter Townsend told Dr. Horton in 1838 that the mine had been wrought 40 rods in length, the general width was 16 feet, consisting of two parallel layers, with a slab of rock between them varying from 4 to 12 inches in thickness. The ore yielded 62 percent iron and the average amount of ore used was 500 tons. This gave 37,500 tons taken from this mine in 75 years. Its iron was remarkably tough, the cost of mining it varied from 50¢ to $1 per ton. The products of the Long mine were cannon-steel, muskets, wire, and fine malleable iron. The iron was also cast into harness buckles, and after annealing, proved exceedingly strong. The iron used for the great chain across the Hudson at West Point during the Revolution was made from equal amounts of ore taken from the Long and Sterling mines, and was smelted at the Sterling furnace during the winter of 1778.

The Long mine was worked as late as 1857. Today the workings are all badly caved-in and flooded.

Morehead or Burrows Mine—The latter name appears on an 1859 map by T. B. Brooks. However, shortly after the Civil War, it was renamed for Joel B. Morehead, one of the directors of the Sterling Iron and Railway Company. Lying about a mile north of Eagle Valley, New York, this mine was worked by means of an open cut and two shafts, one at each end of the mine. The size of the dump indicates that the depth was not great. Workings still visible include an open cut, about 425 feet long, at the north end; a shaft at the south end; and a second open cut about 45 feet in length, 30 feet north of the shaft. The workings cover a distance of about 600 feet and were probably connected underground.

Mountain Mine—In 1838, Peter Townsend, proprietor of the Southfield furnace, told Dr. William Horton, the geologist of Orange County, who was doing a report on the area for New York State, that this mine was discovered in 1758 by a hunter when a tree was uprooted, revealing an iron deposit. The mine yielded 45 percent ore which produced iron remarkable for strength and fine polish. The cost of mining was $1 per ton. There were two dykes crossing the mine at an angle of 45 degrees, each 15 inches thick. Before the Revolution, when this mine was chiefly worked, the iron was sent to England to be used for tinning. Some of the ore was used at the Southfield furnace in 1834–35, but the amount was small compared to the large volume from the Long mine. The Mountain mine lies about a quarter of a mile north-northwest of the Scott-Cook mine's concrete hoisting tower. The workings consist of a series of deep, narrow openings extending 1,100 feet along the strike. This and the Patterson mine can be considered as one unit, as the workings run together. The southernmost openings, on the discovery site, were known as the Patterson mine and the northern openings as the Mountain mine. The mine was probably last worked about 1857.

Professor R. J. Colony, who inspected it in 1921 while doing a report for the State of New York, *The Magnetite Iron Deposits of Southeastern New York,* said, "Nothing can be seen at the present time but an open cut caved in more or less, overgrown with brush, containing water in some places and half full of fallen rock elsewhere." At the north end of the workings a vertical shaft was sunk. The Ramapo Ore Company dewatered this shaft during the course of their preliminary work about 1918 and found it to be 100 feet in depth; a good ore body 7 feet thick was encountered at the bottom of the shaft.

Patterson Mine (This is not the same Patterson mine mentioned by Preston Hotz)—In 1838, Peter Townsend told Dr. Horton that this mine was:

> discovered in 1831 by John Patterson. 1,000 tons of this ore are used annually, which in 7 years gives 7,000 tons as the amount. Cost of min-

ing from 50¢ to $1 per ton; yields 56 per cent. Its ore chiefly used to correct infusible and bad ores, such as O'Neil [sic] and lean hematite ore. Iron good.

Townsend used it at his Southfield furnace in 1834–35 and probably other years too. The Patterson mine in later years practically lost its identity becoming known as the southerly extension of the Mountain mine. The ore body in the mine was about 20 feet thick and had been opened to as much as 150 feet in length.

Redback (***Red or Spruce Swamp***) ***Mine***—Lying just over ½-mile northwest of the Morehead and on the same old wood road, is this mine now called "Redback" after the rusty reddish color of the ore-band outcropping, which still may be traced along the surface. In the earlier days it was known as the Red or Spruce Swamp mine, the latter name coming from the swamp lying just to the south of it. In August, 1939, Robert Henion, gatekeeper at the Eagle Valley Road entrance to Tuxedo Park, when giving directions for getting to the mine, told the author "be sure not to try to cross the old swamp there, as it is all quicksand. A sixty-foot pole would sink out of sight in it." In his younger days Mr. Henion had helped construct the old mining roads through the surrounding mountains, which he said was "very hard work."

In 1838, Peter Townsend noted that the Red or Spruce Swamp mine was "discovered in 1780, by J. Stuperfell; cost of mining 50 cents per ton; ore sulphurous; being remote, not much used; iron sound; has been generally used as a flux, mixed with hard black oxides and refractory cold short ores; it assists fusion, and improves the quality of the iron." [6] At the same time Dr. Horton estimated that the ores found here were magnetic and full of pyrites which would decompose rapidly when dug up and exposed to air. Because of this, he said, the surface of the mine was reduced to powder of an iron-rust color. The ore vein was not more than 10 to 12 feet thick.

Professor R. J. Colony wrote in 1921 that the mine was worked through openings along the strike for 500 feet and the ore was followed down the dip by an inclined shaft, said to be about 300 feet deep, but then inaccessible because of water. He reported that the ruins of the old skipway still remained and parts of the rails could be seen in the slope; and an old circular roaster 20 feet high, rusty and dismantled, stood near the incline which led him to assume that the ore was roasted on the ground.

Today the picture is practically the same, despite the lapse of 40 years. In 1880, the mine yielded 3,638 tons of ore and shipped it by horsedrawn teams to both the Southfield and Sterling furnaces. In fact, ore from this mine was carted to the Southfield furnace until September, 1887, when it was permanently shut down. According to Mr. Henion, the Redback mine was last worked in 1900.

His directions for locating the mine are still good. "Go south on the road from this gate-house until you come to two small bridges at the foot of the hill. Just as you reach the second bridge, an old wood road swings off to the right (west). Follow along this old road, bearing left up over the hill for a mile and a half, passing the Morehead Mine on your way, and you will reach the Redback Mine and its roaster."

Scott Mine—Opened shortly after the close of the Civil War, this was named for Thomas A. Scott, a director of the Sterling Iron and Railway Company. The location of the shaft is easily spotted from the road leading from Route 17 west to Sterling Lake. A tall concrete hoisting tower stands over the opening, just a few feet north of the road. This mine is connected underground with the Cook mine, several hundred yards to the south. When originally opened, it was worked through an inclined shaft, but the Ramapo Ore Company, who leased the property in 1918, sunk a new 600-foot vertical shaft, raising the ore in three-ton skips by means of an electric hoist. The ore was shipped over a narrow-gauge railroad to the main line of the Sterling Mountain Railway in Lakeville, at the foot of Sterling Lake. Any ore needing concentration was carried a few yards north, on an extension of the main line, to the magnetic concentrating mill at the Lake mine. From 1880 to 1887, ore from the Scott mine was carted to the Southfield furnace seven miles away.

The original entrance to the mine lay about 400 feet south of the vertical shaft and its hoisting tower. This old opening, filled in a few years ago and no longer visible, was typical of mines which are dangerous to enter. At the start, the incline of the shaft was very gradual, then after a very short distance it dipped suddenly and plunged down rapidly at an angle of 85 degrees, almost a vertical drop.

Preston Hotz's report on the Sterling mines, published by the Department of the Interior in 1952, gives the following description of the underground layout of the Scott:

> The mine was worked through two levels, 180 and 345 ft below the collar of the vertical shaft in the footwall of the Scott mine. South of the shaft, two large stopes [step-like excavations] nearly 300 ft. high and 500 to 800 ft. long extend almost up to the surface from above

Scott Mine hoist. *Photo by Raymond S. Darrenougué, 1952.*

> the 2nd, or 400 level. The shoots are estimated to have had a thickness of 20 to 40 ft. No mining beyond that necessary for development has been done on the 400 level. In the Scott mine the 200 and 400 levels were driven 1,000 and 1,200 ft., respectively, north of the vertical shaft.

Some writers have erroneously referred to this mine as the Oregon mine. The Oregon was an older mine, almost connected with the Scott, but actually a separate unit. It is shown on an 1854 map of the tract included in a booklet the Townsends published in an attempt to sell the Sterling property.[7]

In one year (1880), the mine yielded 5,783 tons of ore. The Scott mine was finally closed in 1921.

Smith Mine—The shaft of this mine is now flooded and caved-in. It lies just west of the Long mine near a large swamp. It is probably about 30 feet deep. The ore body was said to have been two feet thick.

Steele Mine—Lying about 2,000-feet southwest of the lower Sterling furnace site are a series of pits and trenches that extend about 800 to 900 feet along the strike. These openings comprised the visible evidence of the Steele mine. In a few places where the ore is exposed to view the vein is of poor quality, seldom more than 5 or 6 feet wide. Most of the mining done here has been of an exploratory nature.

Sterling Mine—This well-known mine was discovered in 1750. Peter Townsend told Dr. Horton in 1838,

> by whom unknown; named after Lord Sterling [sic] the then proprietor of the soil; he sold, and a blast furnace was immediately put in operation by Messrs. Ward and Colton, that is, in 1751; cost of mining is 37½ cents per ton. Its yield is always 50 per cent in the blast furnace. Amount of this ore used has ranged from 500 to 2,000 tons annually; the medium of this gives 137,000 tons as the amount of ore used from this mine. At present the amount used is 2,000 tons; the ore always fuses easily; its iron is between cold and hot short; very sound and strong. It has been largely used for casting cannon and for making bar iron; no proper dykes in the mine, it lays on the side of a mountain. The ore, in different places where opened, is from 10 to 20 feet

thick, inclining at an average angle of 30°. The floor is smooth granitic rock, a little over three feet thick; rests on another bed of soft rich ore, and the little used proves free of sulphur. (I may add from my own knowledge *positively,* that another immense bed underlays the last mentioned.—*Wm. Horton.*) Sterling mine covers a surface of more than 30 acres, by survey; part of this the ore is bare, part is covered by soil from one to five feet in depth, and part by rock, from six inches to a yard, or more, in thickness.[8]

At the time of this statement by Peter Townsend, the ore body was clearly exposed in one area about 50 rods wide and 150 yards in length. The mine was worked as late as April of 1902. Shortly afterwards it closed for good.

The entrance to the shaft, now filled in, lies at the southwest corner of Sterling Lake, near the outlet. This location is about 150 feet north of a bend in the road which crosses the railroad tracks at old Lakeville. The shaft passed under a nearby brook[9] and stretched under Sterling Lake for 1,000 feet on a 28 degree slope. The ore vein measured from 10 to 30 feet thick. There is no trace of the mine itself today.

In 1880, this mine, with the Lake and Tip-Top mines, produced a total of 26,173 tons of ore, the greater amount of it coming from the Sterling mine. By this time, all the ore from the immense outcropping had been removed, with further workings being carried on under a massive hanging wall or roof supported by pillars of ore. From 1880 to 1887, the ore from the Sterling mine was used at the Southfield furnace. Although no ore was raised in 1888, the pumps were kept going to have the mine available for use. The slope at that time was 600 feet long according to John Smock, a New York geologist, who submitted a report on the mines in the area in Bulletin No. 7 of the New York State Museum, 1889. Smock pointed out that a large amount of ore was taken from the original outcropping bed on the side of the hill and that the total output of this mine made up a large part of the total ore produced by the company's mines.

Summit Mine—This opening is located in a narrow ravine about 1,500 feet west of the Lake mine shaft opening. An inclined shaft of unknown depth follows the dip of the vein to the east. It is now filled with water. There are many small pits in the area. The size of the ore body is not known. This mine was worked as early as 1857.

Tip-Top or Clarke Mine—The opening is on a hill about 1,000 feet south of the Lake mine. It was never extensively mined, ceasing operation about 1889. The ore was taken out in skip cars through a tunnel in a section of the mine and moved to the bottom of the hill on a gravity-incline railway without machinery and then dumped into pockets of the main line of the Sterling Mountain Railway. Worked from an open cut, the ore is similar to that of the Lake mine.

Whitehead Mine—This mine lies 1,000 feet northeast of the Lower California mine, between the highway and the Sterling Mountain Railway tracks. It is flooded and inaccessible. The ore body, said to be about 8 feet thick, was worked by means of an open cut to an unknown depth.

Interior of Tip-Top Mine showing the gigantic pillars supporting the mine roof. *Photo by Charles F. Marschalek.*

WHYNOCKIE AREA MINES

Beam Mine—Lying three-fourths of a mile southwest of Haskell, New Jersey, in a vale between two knolls, this old mine was reopened in 1875. A small amount of ore was taken out before the mine was abandoned in the same year. A 20-foot deep shaft was worked on a vein 4–5 feet in width. Today two shafts and a few sizeable dumps can be seen.

Blue Mine—Also called the Iron Hill, Whynockie, or London mine, this is a distinctly different opening from its namesake, the Blue mine of the Ringwood group, which lies several miles to the north.

It was probably first known as the London mine, due no doubt to its likely discovery by Peter Hasenclever. It is located about a mile and a half west of Midvale. For a number of years before 1857 it was worked by Peter M. Ryerson, the ore being used at the Freedom furnace until the last blast in 1855. Visitors to the mine in 1858 and 1867 reported it filled with water. At the later date the shaft extended almost 150 feet along the vein. The mine was worked again in 1871 and 1872.

In March 1886, the Whynockie Iron Company under its superintendent, G. M. Miller, reopened the mine, working it over 100 feet down the slope, plus a drift at the bottom 50 feet in length. For a short period it produced about 300 tons of ore per month. Work was halted until 1890, when the mine was again de-watered. After resumption of operations, over 8,000 tons of ore were raised before it was once again shut down. Some years ago a former workman who had helped de-water the mine described one of the steps used in this rather difficult operation.[1] The mine workers, he recalled, stood on a raft which sank lower and lower as the water level declined. They would progressively clear away debris left clinging to the walls of the mine and shore up the timbers in the sides of the shaft as the water was being pumped out. It evidently called for men with good "sea legs" who could keep their balance while working with their hands. Even after another de-watering operation in 1905, the mine was not worked again.

The overall length of the working area measures approximately 500 feet. At its widest part, the ore vein reaches 16 feet in thickness, the average width being from 9–10 feet. The ore was compact, hard and mixed with much rock. Some of the old stonework and concrete foundations can still be seen near the mouth of the mine, scattered at ground level and on the hillside. About 100 yards to the northwest, up a stream, can be seen the remains of what was once a sizeable dam, doubtless used to provide power in earlier days for de-watering the shaft by waterwheel and pump, and later for the mine hoist. The mine is now filled with water. Directions for getting there are under Roomy Mine.

Board Mine—Located on top of a knoll about two miles northeast of old Boardville (a community now submerged beneath the Wanaque Reservoir), this mine is over a mile west of Ringwood. It was opened in 1872. By the end of 1873, the ore, mined and shipped, totaled 11,000 tons. The vein was 9 feet wide and had been worked for a length of 100 feet to a depth of 70 feet. The mine remained idle from 1875 to 1882 and was abandoned late in 1884, after being worked intermittently during the intervening two-year period. Again the depressed state of the iron market caused a stoppage. A second opening made to the northeast of the main one was not worked to any great extent. While the mine was in operation in 1884, the lessee, Joseph L. Cunningham, amassed over 4,000 tons of ore stock on the mine bank. Today all that can be seen is two caved-in pits and a small dump.

Brown Mine—This mine, discovered by S. D. Brown of Paterson, is located in Wanaque, a little over a half-mile southeast of the big reservoir dam. The shaft is 26 feet deep and the vein 12 feet wide. It was opened in 1874, and ore was taken out. Work then stopped and the mine was closed until February of 1880, when it was reopened and worked again for a short period. It yielded about 250 tons of ore per month at peak operation and in 1880 produced a year's total of 1,232 tons. At that time it was operated by the New York and New Jersey Bessemer Ore Company and the ore shipped to the Saucron Iron Company in Hellertown, Pennsylvania.

Butler Mine—Located on the ridge just west of Bear Swamp Lake in the northwest corner of Bergen County, this mine was worked prior to the Civil War and has been the scene of extensive searches for ore. Some good ore was uncovered, but no real mining has ever taken place here. It was first worked for about 20 feet, showing a vein 5 feet thick at the northeast and 8 feet at its southwest end. At a depth of nearly 9 feet, it revealed an ore rich, compact, and free from sulphur. The compass here shows a strong attraction for at least 200 feet to the southwest and lighter attractions for an additional 500 feet.

On lower ground 50 feet southwest of the first opening, a cross cut was made in the earth and dug out to a depth of 7 feet uncovering ore and rock. For a distance of 14 feet across this cut, ore was visible. Nine feet was filled with ore that appeared to be nearly pure, the remaining 5 feet contained a large proportion of rock. Some 70 feet to the southwest is another cut with an ore vein 11 feet thick. Most of the additional exploratory work was done in 1873, 1874, and 1879, but only about 50 tons of ore were removed by R. F. Galloway of Suffern, New York, who worked it then. The mine was reopened in 1880, worked for a month during which it produced 280 tons of ore and then was abandoned. The workings can still be seen but cannot be entered.

Kanouse Mine—Just one-third of a mile east of the main road in Wanaque at the foot of the Ramapo mountains, is this mine with four shafts, plus some surface diggings. The vein, free from rock, is 7 to 8 feet thick. The ore was hard and close grained. The weathered specimens lying in the area are fairly rusty, indicating possible sulphur.

Although opened before the Civil War, the mine was not in operation in 1867. When reopened by a New York Company in 1872, four shafts were sunk and 2,000 tons of ore were extracted by 1875 when the mine was closed. In 1882 further exploratory work was done and a large sum of money was expended to prepare it for operation. Two main shafts 1,500 feet apart, each 100 feet deep, were used to reach drifts with aggregate lengths of 600 feet. Only a few tons of ore were raised before work was stopped again. In 1890 the mine was de-watered and explored, but not worked, as prospects did not appear to justify the expense.

Monks Mine—Lying on the north side of Whynockie Creek about two miles above the site of old Boardville, this mine was opened shortly after the Civil War. The vein was about 7 feet wide, the ore hard and black. The mine was abandoned in 1890.

Roomy Mine—Also called the Laurel or Red mine, it is located three-fourths of a mile northeast of the Blue or Iron Hill mine. It was probably opened shortly after 1840. Extensive operations were carried on here through an adit in the side of a hill, above water level. The passage can still be entered and followed for a distance of about 50 feet. The mine was also worked on the surface along 50–60 feet of its length. The ore was compact and free from rock. The vein, 4 feet thick, had a pitch of 58 degrees, dipping sharply to the southeast. Prior to 1857 about 60 feet of this vein had been worked. This mine is well worth a visit as it can be inspected on the inside in comparative safety. A little further to the south on the same trail is the water-filled shaft of the Blue mine.

To reach these two mines, take the trail south from the corner of Snakeden Road and Ellen Avenue (about two miles west of where West Brook Road crosses the Wanaque Reservoir). It is just a short walk along the trail and most rewarding.

Tellington Mine—Located about four miles northwest of Wanaque are three shafts, the deepest 40 feet. The vein of ore is 5 feet in width. The ore was carted to the Midvale station of the Montclair Railway Company at a low cost, due to the gradual descent of the road from the mine to this point of transportation. The mine was opened by S. D. Brown of Paterson in the spring of 1873. By 1874, it had yielded 200 tons.

Vreeland Farm Mine—This mine, opened about 1878 or 1879, on the farm of Thomas B. Vreeland, close under the cliffs of Kanouse Mountain, is made up of several pits, the deepest of which is 35–40 feet. It lies about one mile north of the old Charlotteburg station.

IV

CHAINS ACROSS THE HUDSON

CHAINS ACROSS THE HUDSON

The opportunity offered by the wide silvery Hudson River for dividing and conquering the Colonies was perceived by both sides even before the Revolution. For Great Britain, with her vastly superior naval forces, a powerful thrust up the river with men-of-war, followed by transports carrying troops to take and hold ground, was an obvious as well as a militarily strategic plan. Washington wrote, "The importance of the Hudson River in the present Contest and the necessity of defending it, are Subjects which have been so frequently and fully discussed and are so well understood that it is unnecessary to enlarge upon them," and he urged "most serious and active Attention to this infinitely important object."

As early as May 25, 1775, the Continental Congress adopted a series of resolutions and sent them to the Provincial Congress of New York in which it was resolved that "a post be . . . taken in the Highlands, on each side of Hudson's River and batteries erected in such a manner as will most effectually prevent any vessel's passing that may be sent to harass the inhabitants on the borders of said river; and that experienced persons be immediately sent to examine said river in order to discover where it would be most advisable and proper to obstruct the navigation."

Five days later, New York's Provincial Congress acted by ordering "That Colo. [James] Clinton and Mr. [Christopher] Tappen be a Committee (and that they take to their assistance such persons as they shall think necessary) to go to the Highlands and view the banks of Hudson's River there; and report to this Congress the most proper place for erecting one or more fortifications and likewise an estimate of the expense that will attend erecting the same."

In their report submitted on June 13, 1775, among other suggestions, the committee recommended that "by means of four or five booms chained together on one side of the river, ready to be drawn across, the passage can be closed up to prevent any vessel passing or repassing."[1] Plans for the fortifications, including the construction of Forts Clinton and Montgomery, erected near the present Bear Mountain Bridge, were started. A security committee composed of John Jay, Robert Yates, Christopher Tappen, and Levi Pawling busied themselves with plans that included various ways of obstructing the river such as *chevaux-de-frise*, fire ships, booms, and chains. Another committee was designated to inquire into such details as finding out about the depth of the water in the Hudson River from Manhattan Island to New Windsor.[2] However, on July 16, 1776, only a few days after the Declaration of Independence had been voted upon by the Continental Congress, the Provincial Convention of New York created the "Secret Committee" whose duties were "to devise and carry into Execution such Measures, as

Pin in the great chain at West Point, New York. *Photo by W. J. Hollingshead, 1953.*

to them shall appear most Effectual for Obstructing the Channel of Hudson's River, or annoying the Enemy's Ships in their Passage up said River; and . . . this Convention Pledge themselves for defraying the Charges incident thereto."[3]

On this newly-appointed Secret Committee were John Jay, Major Christopher Tappen, Robert R. Livingston, and William Paulding. They hurried to assemble three days later—and with good reason. On July 12, only four days before the creation of the committee, the British frigates *Phoenix* and *Rose,* had demonstrated the inefficacy of the river defenses at Fort Lee by sailing past the hastily improvised *chevaux-de-frise* placed in the Hudson between Fort Washington on the east shore and Fort Lee on the west shore. The ease with which these enemy ships passed both the incompleted underwater barrier and the combined cannon fire of the twin forts alarmed the New York committee.[4] And then, as if to prove it was no trick at all, the *Phoenix* and the *Rose,* after lying at a mooring in the Tappan Zee above Fort Lee and Fort Washington, sailed back down the river, again running the gauntlet without serious damage and demonstrating their ability to withstand the river defenses there. This episode demonstrated the need for completing the river obstructions as soon as possible. The *chevaux-de-frise* was formed of stone-loaded log cribs holding projecting iron spars, and was sunk in the river channel as a barrier. Sunken ships were also used.

The final test of the obstruction soon came. On October 9, 1776, Tench Tilghman described the event in his letter to the Convention of New York:

> About 8 o'Clock this morning the Roebuck and Phoenix, and a Frigate of about 20 Guns, got under way from about Bloomingdale, where they have been laying for some Time, and Steered on with an easy Southerly Breeze towards our *Chevaux-de-Frize,* which we hoped would have given them some Interruption, while our Batteries played upon them; but to our Surprise and Mortification, they all ran through without the least Difficulty, and without receiving the least apparent Damage. How far they intend to go up, I don't know; but His excellency thought fit to give you the earliest Information, that you may put Gen. Clinton on his guard at the Highlands; for they may have Troops on Board to surprise those Forts.

It was a big disappointment to see the *chevaux-de-frise* fail. The highland forts up the river were only partially completed.

On November 18, 1776, William Heath (in command of the Hudson Highlands area) wrote to James Clinton:

> I have just Received the Disagreeable News of the Reduction of Fort Washington, the day before yesterday, but have not as yet had the Particulars. The Enemy will now have Possession of the River below us, and it Behooves us to Exert every Power to render the River Impassable at the Highlands. I beg therefore that you would Endeavour to Compleat your works and the obstruction in the River as fast as Possible.

Orders were given to improve the obstructions, but when Fort Washington fell to the British on November 16, and Fort Lee shortly afterward, little had been accomplished. The *chevaux-de-frise* there was abandoned and work was speeded up in the Highlands, for it had been proposed and agreed to by the Secret Committee at a meeting at Fort Montgomery on July 19:

> that in order to Obstruct the Navigation of the Hudson's River so as to prevent any of the Ships of the King of Great Britain coming up the same, it will be necessary to throw across the River at or near Fort Montgomery, a Boom, and below it to Anchor Frames of Timber, the Points or Ends whereof to be shod with Iron, so as to answer the double purpose of Pounding any Ships that may Sail up to it; and if that should fail, to lessen the shock of those vessels when they come to the Boom—such Frames to be made in the following Manner: The pointed Beams to be of about the length of 16 feet and to be about 15 foot apart and two cross Beams worked in and Bolted.

Six days later, at Poughkeepsie, the same committee recommended putting a chain across the Hudson and hastily moved to put this into effect by agreeing:

> that an express be sent to Gen. Schuyler for the chain intended to be thrown across the River Sorel, to be employed for the above purpose; and as it may fall short of the distance required, it is farther concluded to apply to Col. R. Livingston to make, until countermanded by this committee, a quantity of bar-iron of about 1½ inches square and to be sent from time to time to the works at Poughkeepsie.[5]

Full details concerning the Sorel chain have not been found. According to a letter from General George Washington to General Philip Schuyler at Albany, it had apparently been ordered by General Charles Lee, perhaps at the request of the Continental Congress. Washington advised Schuyler that "you will also receive the Chain which General Lee order'd and which I think should be sent to and fix'd at the Place it is designed for, with all possible Expedition." In another letter written at the same time to John Hancock, President of the Continental Congress, Washington related that "I have sent with the last Brigade [to Canada in May of 1776] . . . also the Chain for a boom at the Narrows of Richelieu [flows north from New York State line to Sorel in Canada] . . . and have wrote General Schuyler to have the Boom fixed, as soon as possible." [6] The threat posed by British forces in Canada, including their naval forces, of mounting an attack on Lake Champlain and Fort Ticonderoga evidently was responsible for having the Sorel chain made and transported to its proposed site.

Whether this chain was found to be ineffective or, more likely, failed to be placed in position due to the hurried retreat of the Revolutionary forces from Canada, is not known. But it was carried down to Fort Ticonderoga, with the planned objective of stretching it across the narrow part of Lake Champlain between Ticonderoga and the camp of the Continental forces on the eastern shore. On July 25, 1776, the Secret Committee wrote to General Schuyler:

> as the chain intended to Obstruct the River Sorel cannot now be applied to that Use and will Serve to prevent the Enemy's Ships from going beyond the Hook on Hudson's River, we must beg the Favour of you to send it (the whole or such Parts of it as may expeditiously be had) to Poughkeepsie and consigned to Messrs. Van Zante, Lawrence and Tudor with the utmost dispatch. Be pleased to inform these Gentlemen of the length of such Part of the Chain as you can send in order that they may direct the defficiency [sic] to be supplied. . . .

Schuyler's reply expressed doubt that the chain could be spared, but if not, "Messrs. Van Zante, Lawrence and Tudor will be Advised of it without delay." [7] Either Schuyler later changed his mind or, more likely, he was overruled by the Secret Committee. On August 13, the Committee drew up "an order in favor of Dirk Schuyler for 15 pounds for bringing the Chain and some Pine Knots from Albany."[8] On the same day, Robert Yates, one of the Committee of Defense, wrote to Washington that:

> the chain intended for the Sorel is arrived and will form a quarter part of the one designed for Hudson's River. The iron for the remainder is come to hand and the smiths began this day to forge it. We have agreed to fix one end of it at Fort Montgomery and the other at the foot of a mountain called Anthony's Nose. It will cross the river obliquely, and for that reason be less exposed to the force of the tide and less liable to injury from the ships of the enemy. The length of the chain will, at least, be twenty one hundred feet.[9]

When the chain was first placed in position across the Hudson during the early days of November, 1776, it failed to live up to expectations. Pierre Van Cortlandt, President of the New York Convention, wrote to John Hancock, President of the Continental Congress about the frustrations which beset the planners. "The great length of the Chain, being upwards of 1800 feet, the Bulk of the Logs which was necessary to support it, the immense weight of the Water which it accumulated, have baffled all our Efforts. It separated twice after holding only a few Hours." This failure momentarily dimmed the hopes of everyone for obstructing the river with a practical barrier.

On December 9, 1776, the Secret Committee met to hear a report on the workmanship of the chain which was believed to have been faulty. General James Clinton; Captains Abraham Swartout and James Rosencrans; and Lieut. Daniel Lawrence placed the blame on two weak parts "a Swivel broke, which came from Ticonderogo" and "a Clevis broke which was made at Poughkeepsie" and added that "no flaw [was] to be seen in any Part of said Chain." The chain was taken up from the river and sent to Samuel Brewster's Forge at New Windsor, New York, where necessary repairs were made during the winter of 1776–1777. By March 14, 1777, Governor George Clinton reported that replacing the chain in the river awaited only the anchors and cables used for drawing it across from shore to shore. It was finally put into place at Fort Montgomery about ten days later.

In letters written from his headquarters at Morristown, New Jersey, Washington displayed his interest in the fate of the obstruction and its probable effectiveness. On April 20, he expressed his

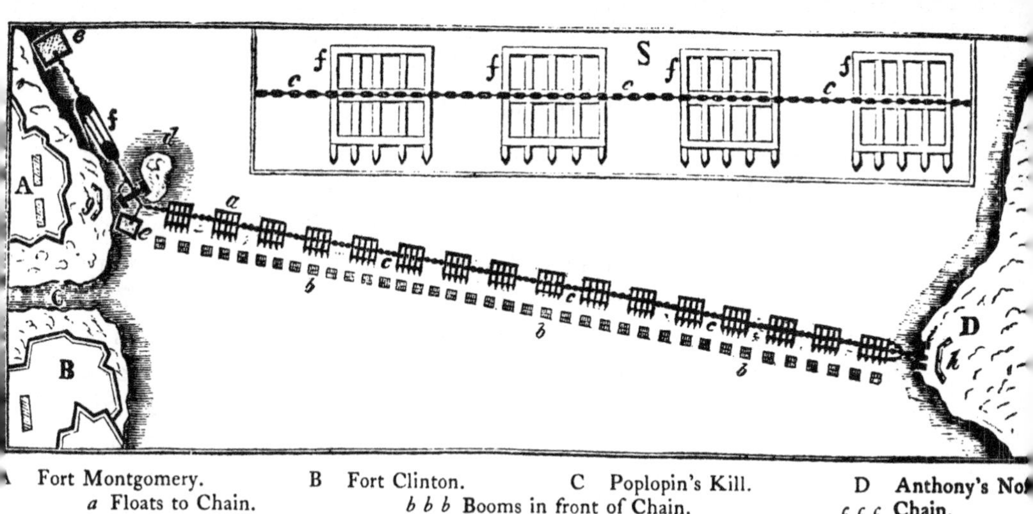

Fort Montgomery chain and boom. From the Papers of the Secret Committee, E. M. Ruttenber, *Obstructions to the Navigation of Hudson's River*, 1860.

pleasure "to hear that the Chain across the River promises to answer the end." [10] Little over a month later, writing to the President of the Continental Congress at Philadelphia, he pointed out that "the cables must be proportioned to the width of the River which is 540 yards . . . a boom across the river with one or two cables in front, if diagonally laid, the gentlemen think they shall not be less than 459 fathoms long and of the largest size that can be had. Unless they are large and substantial they will answer no purpose and will not sustain their weight when stretched." [11] The Commander of the Continental Army was eager to contribute whatever he could in the way of experience and advice. Great hopes were held for the Fort Montgomery chain and its defense against British river-borne attacks. Six months later its use was over—without its ever being put to a final test.

"On the 6th of October at the break of day [I] landed the troops at Stony Point." Thus Sir Henry Clinton described the start of his attack on the twin forts across the river from Anthony's Nose. Approaching Forts Montgomery and Clinton from the west or opposite side, his force of some 2,000 men stormed into the forts and won a resounding victory, completely outflanking the great chain on the Hudson River, and routing the relatively smaller number of Continental troops defending the forts. Whether Clinton purposely avoided testing the chain on the river or merely chose to attack the rear of the fortifications as a sound military tactic is problematical. But his strategy might easily have included both. At any rate, he was thoroughly aware of the presence of the Fort Montgomery chain, as it was called. In his *Narrative of His Campaigns 1775–1782*, Sir Henry Clinton mentions destroying the various kinds of works at the forts and specifically states that "as soon as they [British forces] had removed the small chain which was run across the river, Sir James [Wallace] proceeded to explore the cheveau-de-frise between Polipals Island and the mainland and everything in the river beyond it." A British historian has said that "Every article belonging to their laboratory, which was in the greatest perfection. . . . the boom and chain which ran across the river from Fort Montgomery to St. Anthony's Nose, and which is supposed to have cost £70,000 fell into the hands of the conquerors." In a footnote he states, "This chain was of most excellent workmanship; it was sent to England from there to Gibraltar, where it was of great use in protecting the shipping at the moles [breakwater]." [12]

Many writers in the past have believed that the Ringwood ironworks played a part in the making of the Fort Montgomery chain. There is a bill written by Erskine in the museum adjacent to Hasbrouck House (Washington's headquarters at Newburgh, New York) made out to Thomas Machin and John Nicoll for the articles of iron manufactured from August 21, to November 1, 1777, which seems to corroborate this belief.[13] Another, from Noble and Townsend, proprietors of the Sterling works, to the same two representatives for 142 Clips, 58 Chains, 21 Swivels, 7 Clevises, 184 Bolts and 8 Bands, seems to bear out the contention that the Sterling works, too, had a part in the making of this iron river barrier. The great chain and its boom were both, however, already in place across the river when the first loads of these products from Ringwood and Sterling were delivered to the Hudson River. These were evidently boom iron pieces ordered from Ringwood and Sterling as substitutes for the rafts of timber which had been placed in front of the chain. Before any decision could be made as to whether the timber rafts should be taken up and replaced by the iron boom, the British attack ended any hope of carrying out this idea and so there is still no actual proof that any iron produced at the two famous furnaces and forges of the Ramapos was ever used in the Fort Montgomery chain or boom.[14]

The inadequacy of the fortifications, including the chain, was quickly realized by the Patriots. In order to remedy this and make the Hudson and its adjoining areas, particularly West Point, less vulnerable to the British, Washington placed General Israel Putnam in charge, instructing him to consult with Governor George Clinton, General Samuel H. Parsons, and the French engineer, Colonel Bailleul de Radière. Putnam, in turn, sought the advice of the New York Convention which eventually appointed five commissioners. A report submitted January 14, 1778, by these commissioners assessed the natural features and probable measures that would be needed to block as many of the routes leading to West Point as possible, stating that

> there are so many Passes across the Mountains to this Place, that it will be almost impossible for the Enemy to prevent the Militia from coming to the relief of the Garrison [at West Point]. By bottling up the Hudson effectively, one main route would be denied to any attack . . . the Committee are led to conclude that the most proper place to obstruct the Navigation of the River is at West Point; but are at the same Time fully convinced that no Obstruction on the banks of the River can effectually secure the Country, unless a Body of light Troops, to consist of at least two thousand effective Men, be constantly stationed in the Mountains while the Navigation of the River is Practicable, to Obstruct the Enemy in their approach by land.[15]

One immediate result of this comprehensive report was the visit of Captain Thomas Machin, an outstanding engineer, to the home in Chester, New York, of Peter Townsend, one of the proprietors of the Sterling works. There he discussed a contract to turn out the chain which would be needed to block the Hudson at West Point. A few weeks following Machin's trip to Chester, Hugh Hughes, the Deputy Quartermaster General of the American Army, arrived to conclude the contract (February 2, 1778), initiated by Machin for the production of the West Point chain. No time was lost in starting work on the links and other parts the following day. The evening after Hughes negotiated the contract with Peter Townsend, he wrote an account of the circumstances and the conditions under which the contract was signed.[16] Among the stipulations made by Townsend was that the Sterling teams should be exempt from hauling military supplies for nine months. Hughes regretted this, but realized the necessity. However he did limit the number of teams.

Peter Townsend II's version of the meeting when the contract was signed is contained in a letter which he wrote March 10, 1845, when he was seventy-five. Consequently, the facts became rather twisted due to the long lapse of time between the event and its retelling. After briefly setting the scene when the Fort Montgomery chain was placed across the Hudson, he claims that this chain "had been made at Ringwood furnace, New Jersey, was of small diameter and composed of cold short iron of an inferior quality and upon the surrender of the Fort fell into the hands of the British." [17] Townsend continues with the decision of the "Council" [committee] to fortify the river at West Point:

> To effect this object, it was determined, among other things, that a chain should be thrown across the River, the links of which were to be double the diameter of those in the chain used at Fort Montgomery, and that it should be constructed of the very best iron the country afforded, and be capable of resisting any force that might be brought against it.

> My father, Mr. Peter Townsend, of Chester, Orange County, was at this time the owner of the Stirling Iron Works, situate at Stirling, in said county, in the mountains at the distance of some 25 miles back from West Point. . . . Application was made by Colonel Pickering to make the chain in question. I distinctly remember the arrival of Colonel Pickering at my father's house in Chester late on a Saturday evening, in the fore part of March, 1778.[18]

> His plans were at once warmly entered into by Mr. Townsend, and such was the ardor of the Whigs of those days that both Gentlemen left Chester at midnight in the midst of a violent snow-storm and rode over to the Stirling Works, a distance of fourteen miles, to take measures for commencing the work. At daylight on Sunday morning Mr. Townsend had all his forges in operation, and his patriotic workmen engaged upon the chain. The work was prosecuted day and night without interruption until its completion and was finished in six weeks.

> It [the chain] weighed 140 to 150 tons, was of unsurpassed quality of Stirling Iron and of superior workmanship. It was carted to the

> River by New England teamsters in sections as the same were from time to time completed.

Although past writers have often cited Townsend's letter as the authority for details connected with the manufacture of the famous West Point chain, other sources disagreed—and rightly so—with some of Townsend's statements. He apparently relied, to a great degree, on his memory of the events centered around the meeting of his father with Machin and Hughes, the subsequent results and details. These had all occurred at least sixty-five years before he wrote the letter. A boy of only seven or eight at the time, Townsend's recollection of the persons and events suffered from the passage of years. His first-hand account as a virtual on-the-scene observer, though possessing some merit, is not wholly reliable.

The ore for the West Point chain was furnished by the Long and Sterling mines. As the links were finished, they were hauled by oxcart in groups of ten to Captain Machin's forge at New Windsor for forging together, before being floated down the river to West Point on logs. During production it was realized that extra strain would be imposed on the chain at certain points: some links were made stronger and heavier than others, becoming as much as three and a half inches thick and as long as three and a half feet. Work on the chain was carried on at Sterling twenty-four hours a day and the Continental Army even furloughed sixty men to speed the operation. In a progress report to General Washington, February 13, 1778, General Putnam wrote:

> At my Request the Legislature of this State have appointed a Committee to affix the Places and manner of securing the River, and to afford some Assistance in expediting the Work. The State of Affairs now at this Post, you will observe is as follows: The Chain and necessary Anchor are Contracted for, to be completed by the first of April; and from the Intelligence I have received, I have reason to believe, they will be completed by that Time. Parts of the Boom intended to have been used at Fort Montgomery, sufficient for this Place are remaining. Some of the Iron is exceedingly bad, this I hope to have replaced with good Iron soon . . .[19]

The importance of this letter lies not only in the disclosure of the date set for completion of the West Point chain, but also in revealing clearly that boom iron made at Ringwood for the ill-fated Fort Montgomery barrier arrived too late to be used and would form a part of the second Hudson River chain and boom.

It is possible to follow some of the progress of the West Point chain by reading the "expense account" of Captain Machin who was the key man in this project: "Jan. 16 Expences on the Road to Chester to Agree for the New Chain, 3 days . . . Jan. 20 Expences Getting Timber for the Chain four days . . . Jan. 26 Getting up Drift Timber . . . Mar. 5 Getting logs to Drye for the Chain at New Paltz . . . Mar. 29 Expences to West Point . . . Apr. 7 Expences Getting Down the Chain Logs with 40 Men, 4 Days . . . Apr. 16 Taking Down the Chain" and finally, "Apr. 30 While Getting the New Chain Across."[20] The speed with which the task was accomplished, is a tribute to the untiring attention and concentration of Machin. Estimates of the time in which the chain was made vary, according to different sources, from six weeks to eleven weeks.[21] Machin's expense sheet indicates that it must have been finished sometime before April 16, the day on which he records "Taking Down the Chain," presumably from New Windsor to West Point. Of course, this does not give any clue to the actual date when the Sterling works completed its project of forging the links into final form. A letter written to Machin from William Hawxhurst, agent for Noble, Townsend & Company, April 23 reports: "I am just now from Nobels at Sterling. The Chane is going on fast" and seemingly contradicts Machin's brief comment about delivering the chain. Various other sources give differing information. Perhaps the most exact is a letter from Machin to General Alexander McDougall on April 20 in which Machin reports that "Lieut. Woodward, who I told you was at Sterling Iron Works inspecting the chain is now returned and informs me that 1700 feet of the Great Chain, which is more than equal to the breadth of the river at the place last fixed upon, is now ready for use. The capson [capstan] and docks are set up at the lower place and the mud blocks are launched, and only wait for good weather to carry them down . . . if the weather should be favorable I am in hopes we shall be able to take the chain down, all fixed in about six days." This communication, on the other hand, is obviously at odds with Machin's own journal of expenses incurred on the job. At any rate nearly three months elapsed from

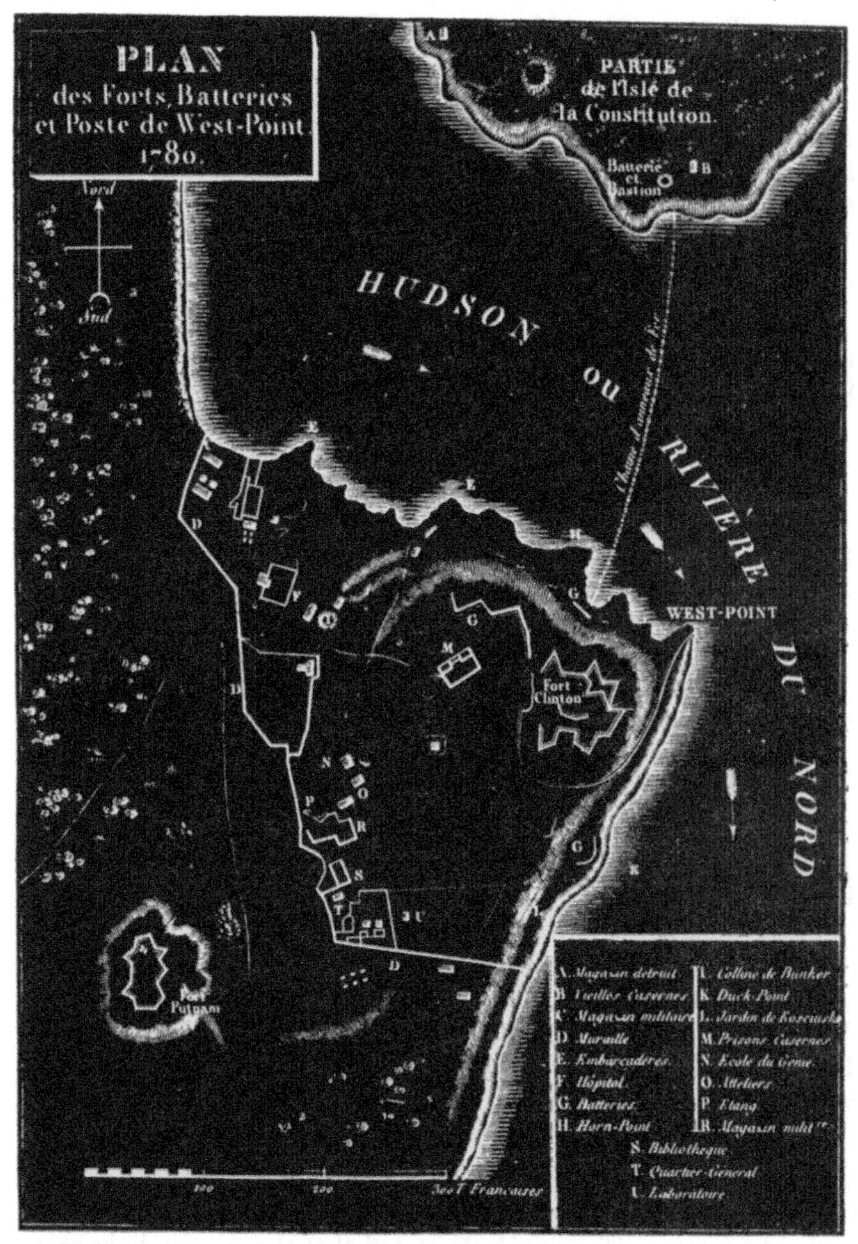

Old French map. Barbé-Marbois, *Complot D'Arnold et De Sir Henry Clinton,* 1816.

the time Townsend took the order until the chain was placed across the river at West Point on April 30, 1778.

Numerous descriptions of the chain still exist. In his *Military Journal,* Surgeon James Thacher pictures it as "buoyed up by very large Logs of about sixteen feet long, pointed at the Ends, to lessen their opposition to the Force of the Current at Flood and Ebb Tide. The Logs are placed at short distances from each other, the Chain carried over them and made fast to each by Staples."

Deducing the boom's probable construction from fragments and pieces recovered from the Hudson in 1855, E. M. Ruttenber wrote in 1860:

> The Relick here referred to consists of two Logs, one of White Wood and the other of White Pine, about fifteen feet in length and about twelve inches in Diameter, dressed in the Center in the Form of an Octagon, and rounded at the Ends. These Logs are united to each other by an Iron Band around each End and two Links of Chain of nearly two inch Bar Iron, but which have evidently lost much of their original Size from Corrosion [over 75 years in the Hudson]. This Boom extended the whole Width of the River, in Front of the Chain. It will be observed that the Boom combined great Strength with Practicability. It was, indeed, the Main Obstruction and was placed in Front of the Chain to receive the full force of approaching Vessels.[22]

A part of it can be seen in the museum at Washington's Headquarters, Newburgh, New York. The contract specifications are, of course, the most authentic of all. There has always been a vast difference of opinion regarding the overall weight of the complete chain. Estimates range from a low of 35.2 tons to as high as 186 tons. The latter figure was given by Peter Townsend to Dr. William Horton in 1838, although in a later letter of Townsend's he gave the weight as from 140 to 150 tons as already noted. Other writers claim the chain weighed about 136 tons but Simms and Harte both agree that the tonnage was considerably less, Simms assigning it a figure of 50 tons "outside the Anchorage" while Harte suggested it was probably not over 60 tons. Harte stated that the smaller links were two feet in length, made of 2.25-inch bar, weighing approximately 37 pounds per lineal foot made from a rough squarish bar. The larger links, made of 3.5-inch chamfered bar, were 44 inches long and weighed 83.7 pounds per lineal foot. There was a clevis for every

1,000 pounds of weight and a swivel for every 100 feet of length. Harte also declared that the boom had two lines of connecting chain of lighter weight but the fact that the logs were only ten feet apart made the use of swivels unnecessary; use of shackling straps on the boom made the need for clevises superfluous. The anchorage system used to moor the great chain required, according to the original contract, 12 tons of anchors, a sizeable amount of tonnage. Twelve links of this chain, together with a swivel and a clevis, can be seen today on the grounds of the United States Military Academy at West Point.

A number of legendary and factual anecdotes are linked with this historic chain. In *Narrative of Some of the Adventures, Dangers and Sufferings of a Revolutionary Soldier* (1830), a running account of a veteran who had been stationed at West Point in 1783, Joseph Plumb Martin says, "the soldiers used to denominate [it] General Washington's watch chain." This phrase was undoubtedly a pun on the word "watch" and Martin adds that "the putting down or the

Section of the great chain at the United States Military Academy, West Point. *Photo by A. A. Kepler.*

keeping up of the chain was the criterion by which we were to judge of war or peace." A more sinister legend connects the traitorous activities of Benedict Arnold with an attempt to weaken it so that British seizure of West Point could be accomplished more easily. Arnold is said to have told Clinton that he would have a link removed, ostensibly to have repairs made to it, thus weakening the barrier. Actually, only one reference to the chain could be found in the secret correspondence between Arnold and the British. This is in a letter dated June 16, 1780, in which the American turncoat takes a rather dim view of the chain's effectiveness. "I am convinced the Boom or chain thrown across the River to stop the Shipping cannot be depended on. A single ship, large and heavyloaded with a strong wind and tide would break the chain." [23] Governor George Clinton is said to have with others, "walked across the River on the Chain" according to a statement made by James Wood many years ago to the author of *The Frontiersmen of New York,* Jeptha Simms. Wood said he had also walked across on the chain. According to the historian Ruttenber, the boom could have been readily converted into a bridge capable of allowing troops to cross from one side of the Hudson to the other.

The condition of the West Point chain was a matter of concern that cropped up in correspondence at times during the late summer and fall of 1780. It was reported to be sinking at various spots along its length due to the water-soaked condition of the logs which supported it. On November 19, 1780, Hugh Hughes, the deputy quartermaster general, mentioned matters concerned with "resurrection of the Chain" in a letter to a Major McCarthy. The sodden, faulty logs seem to have been replaced from time to time with fresh, buoyant ones, a task requiring vigilant attention to keep it effective and in proper condition to fulfill its role. Every year at winter's approach, the chain, boom and floats, were pulled up on shore; in spring they were put back again. Although it was in constant readiness, the mammoth chain was never tested in action. It is still a matter of conjecture whether this mighty iron barrier would have been able to measure up to the high hopes held by the civilian and military planners.

Just what happened to the West Point chain after the war is somewhat confusing, yet fascinating. One section of the chain can be clearly and definitely located. This is the portion at the Military

West Point, as seen in the fall of 1778.

Explanation.—A, a battery on Constitution island. B, the great chain suspended across the Hudson. C, Fort Clinton on the West Point. The latter, which occupied nearly the present site of the *Military Academy*, commanded a southern approach to the Point.

West Point, an old engraving. Jeptha R. Sims, *The Frontiersmen of New York,* 1882.

Academy—as noted previously. Another part of it was sent to the West Point foundry at Cold Spring, New York, where it was melted and used for other purposes. A third portion which had sunk to the bottom of the Hudson was recovered years later. From time to time, tales have been told of other links or parts which have turned up in museums or in the homes of individuals but most of these have never been clearly identified as coming from the West Point chain.

An ironmonger said to have been named Westminster Abbey is supposed to have acquired some of the parts at an auction held at the Brooklyn Navy Yard on August 30 and 31, 1887, but the factual details are unclear and tend to be unbelievable.[24] However, Mr. Abbey advertised his newly-purchased historical links for sale, claiming that they were part of the great West Point chain and that he had proof of their authenticity. He found at least one ready customer in Abram Hewitt of the Ringwood ironworks.[25] However, Hewitt, not entirely convinced, wrote to the Navy Department, asking their assistance in substantiating the validity of his newly-purchased items only to be told that the Abbey chain's history could be traced back no farther than fifty years and that beyond that time it was impossible to determine its background.

About a year later Hewitt expressed extreme doubt about the history of the Abbey chain after comparing it with genuine parts of the Hudson River one.

> I purchased from Mr. Westminster Abbey, 36 links of a chain which at the time was believed to be a portion of the obstruction which was laid across the Hudson River at West Point in 1777–1778.[26] Mr. Abbey assured me he had purchased it at a sale at the Brooklyn Navy Yard and had discovered subsequently it was the West Point Chain. After the links were delivered at Ringwood, I found that the dimensions were greater than those attributed to the Chain at West Point. The portion of the Chain at West Point in a trophy, are about two feet long and weigh 140 pounds. The chain I purchased has links of three feet in length, weighing 300 pounds. . . .

To his credit Hewitt never hid the fact that he had bought "a pig in a poke" and even informed his friends of the fraudulent chain. Despite this, MacGrane Coxe, a relative of the Townsend family through marriage, who lived at the old Sterling Manor House in Southfield, New York, apparently sought Hewitt's help in obtaining some of the Abbey chain links. Hewitt's letter to Coxe, however, accepted at face value the request to buy the spurious links while warning Coxe to "let the buyer beware."

> On reflection I think I can arrange for the chain you want, much better than to apply to Mr. Westminster Abbey. I have already notified him that he has sold the chain to me on false representation and I think it would be very unpleasant to apply to him for more of it under the circumstances. I find I have two lengths at Ringwood, each containing eighteen links. Both lengths were laid upon the lawn for experiment. I find that one of the lengths is sufficient for my purpose. I can therefore sell you the other. My cost as I told you was eleven dollars a link in New York and the expense of delivering it to Ringwood has been about a dollar a link additional. My man tells me he

> can deliver one of these lengths of the chain at Southfield for about ten dollars, thus avoiding extra handling, as well as the Railroad freight. He can haul it with four horses from Ringwood to Southfield in the course of a day. To get it up the hill to your house would require an extra team or so, which you doubtless can supply. I therefore suggest that I deliver you eighteen links at the rate of twelve dollars per link at Southfield, which I think will reimburse me for the outlay I have made.

Even after reading Hewitt's letter, Coxe wanted part of the chain and purchased ten links from him.[27] The twenty-six remaining links can still be seen at Ringwood on the lawn in front of the Manor House.

Any hope that Hewitt may have had about finding some small shred of evidence that would prove the Abbey chain to have had a connection with the Hudson River chain at West Point must have vanished at the beginning of the twentieth century. On January 10, 1901, after what may have been a rather long and exhausting search to establish the true identity of the Abbey chain, Hewitt wrote to the ironmonger with the distinguished British name:

Chain links at Ringwood, once owned by Abram Hewitt.

> I am now fully satisfied that it never was and never could have been any portion of the Chain stretched across the River during the Revolution. There were three chains,[28] the largest of which was the third one. The links weighed about 140 lbs. and they were made from 2½ inch iron. The chain which you sold me is made from 3½ inch iron and the links weigh 300 lbs. On further inquiry I have ascertained that it was simply what is termed dredging or buoy chain used by the Government for holding vessels above the buoys and was sold for want of further use. You are bound to take back the chain which you have sold to me, with all expenses incurred thereon, and to return to me the money which I have paid. . . .[29]

An entirely different point of view regarding the chain that turned up at the Brooklyn Navy Yard is reflected by C. B. F. Young, an instructor at Columbia University, in his article "The Great Iron Chain" which appeared in the June 3, 1937 issue of "The Iron Age" magazine. Young mentions a John C. Abbey rather than a Westminster Abbey in his writing:

> A portion of the chain (that which was recovered from the bottom of the river) stayed at West Point for some time, and, during the Great Sanitary Fair of the Civil War held in New York City, a portion of it was sent down for exhibition purposes, with the understanding it was to be returned to West Point. However, the managers, on account of its great weight, found it much cheaper to send the chain to the Brooklyn Navy Yard.
>
> This part of the chain stayed at Brooklyn until 1887 when the United States Navy Department began consolidating the different bureaus under one head called the General Storekeepers Bureau. Commander R. W. Mead was appointed president of a board which was to visit all Navy Yards and examine all stores and anything not in use or fit for Navy service was to be sold at public auction. The old chain was located, and, inasmuch as none of the officers ever heard of the blockades across the Hudson River during the Revolution, they thought it was junk and sold it as such at an auction Sept. 4, 1887. W. J. Bannerman & Co. was the purchaser and in turn sold it with a lot of other scrap to a forge company.
>
> Later two men, F. C. Gunther and M. J. Savage, were looking for relics for the Libbey Prison museum and began to trace the whereabouts of the historic chain. John C. Abbey, a buyer at government auctions, heard about these two men wanting a certain chain and purchased the historic relic from the forge company. Mr. Abbey controlled

the chain for ten years, selling some links to collectors. Later Francis Bannerman repurchased the remainder of the chain and cut most of it up into small parts for use as paper weights, and has sold a number of these souvenirs.

Also, sections of the chain have been sold to various people around the country. Over 20 links are at Ringwood Manor about 30 miles from New York City. The author knowing this, contacted the owner Mr. Hewitt, and asked for permission to secure a section of one of these links so as to make a thorough study of the iron. This permission was refused. However, after tracing down several pieces of the chain it was discovered that Francis Bannerman & Sons, New York City, had two small pieces left of their supply. This company was quite willing to cooperate and allowed the use of one of these pieces for a suitable period. The piece could not be cut in any way, but etching the whole surface was permitted.

Professor Young goes on to describe a technical analysis then made of the piece of the link, but makes no mention of any doubt of its authenticity as a part of the West Point Chain. However, C. Rufus Harte, in his *The River Obstructions of the Revolutionary War*, has this to say:

> The many hands through which the pieces of the chain passed after the sale in 1887, gave ample opportunity for the "forgery" which some "doubting Thomases" claim is the explanation of the marked difference in the appearance of some of the links, namely, that some one, having in mind the good market for the relics, had a considerable number of links made for him. The writer has only hearsay evidence as to what is said to have occurred, but the appearance of some of the large links is certainly a bit suspicious.

A discussion of the pros and cons will probably go on for years to come, but no matter how often the question of proof arises concerning the chain's links, nothing can ever detract from the colorful role the great West Point chain played in America's history.

V

THE CANNON BALL ROAD

THE CANNON BALL ROAD

Meandering across the ridge of the Ramapo Mountains on a generally northeastern course from Pompton, New Jersey, to Stony Point, New York, on the Hudson, is a series of half-obliterated trails and old wood roads. This is the part-legendary, part-historic route called the Cannon Ball Road and exists primarily in the misty folklore of the foothill country of the Ramapos. Its story is based upon family tales of the region and a few known details of the situation during the Revolution.

A much older road—the main one—through the rolling valley east of the Ramapos connected West Point and Philadelphia by way of Morristown and Trenton. Throughout its history, this strategic road has had several names. Some referred to it as the "Valley Road"; others called it the "road to Suffran's"—in reference to the settlement near the mouth of the Clove where John Suffern, ironmaster and Patriot, kept a tavern. The names themselves are an indication of the road's age. Today it is designated as Route 202. It parallels the Ramapo River in its course from Suffern to Pompton and was probably once an Indian trail. However, with the outbreak of the Revolution, the road reached new importance as a main line for the Continental Army and its supply wagons. Along its length were scattered the well-kept plantations of the prosperously economical Dutch and English farmers.

By the time the Colonies had begun active resistance to the British, Tory and Patriot sympathies were dividing formerly friendly neighbors along this valley road. As a highway for military supplies to the Continental Army it invited retaliation from British forces who were kept informed by Loyalist sympathizers living there. Consequently, an alternative road is said to have been built along the top of the Ramapos where the woods concealed the journeying of the wagons with their loads of cannonballs and other vital materials.

Evidence of the actual existence of the Cannon Ball Road is extremely limited. As late as 1892, the opening words of a petition for the residents of Pompton Lakes, New Jersey (the western end of the road), give some credence to its reality: "Beginning at an iron stake in the middle of the Old Cannon Ball Road. . . ."

Maps of the Ramsey Quadrangle issued by the U.S. Department of the Interior in its Geological Survey Map Series (since 1943) label certain trails and roads in the Ramapos north of Oakland as the "Cannonball Trail." Another map, old and undated, has just come to the author's attention. It is an interesting and valuable survey of the Rotten Pond area. But these maps are the only documentary sources and are far from conclusive.[1]

Not all of the meager evidence about the Cannon Ball Road is in writing, and it exists as a firmly established fact in the still-alert mind of Mrs. Susan Roome Horton, an octogenarian of Pompton Lakes. A direct descendent of a local Revolutionary militia officer, Major Adrian Post, Mrs. Horton clearly recalls the story which has been handed down in her family from generation to generation for almost two centuries. In her account, Mrs. Horton says that the road took its name from an incident when a group of Hessian soldiers up-ended a wagon load of cannonballs on the Pompton end of the trail. She does not know the exact year of this incident, but is certain of the actuality of the road.[2]

In 1922, Major W. D. Ennis, an official of the Green Mountain Hiking Club and an engineer by profession, wrote several articles for the *New York Evening Post* describing his experiences in tracing this almost-unrecognizable trail and his conclusions.[3] He illustrated one article with a detailed map of its southern portion from Pompton to Rotten Pond (now Ramapo Lake) and another with a map of the Ramapo ridge north to Hillburn. Ennis seems to have been a competent and careful observer. He spent considerable time inspecting parts of what he believed was the Cannon Ball Road and concluded that it was a well-laid-out military route. He cited apparent workman-like construction in certain places for by-passing marshy ground; the manner the trail wove around hills and rocks, and the use of solid fill or ballasting in other places for leveling the roadbed. An earlier investigator, Frank Place, a co-author of the

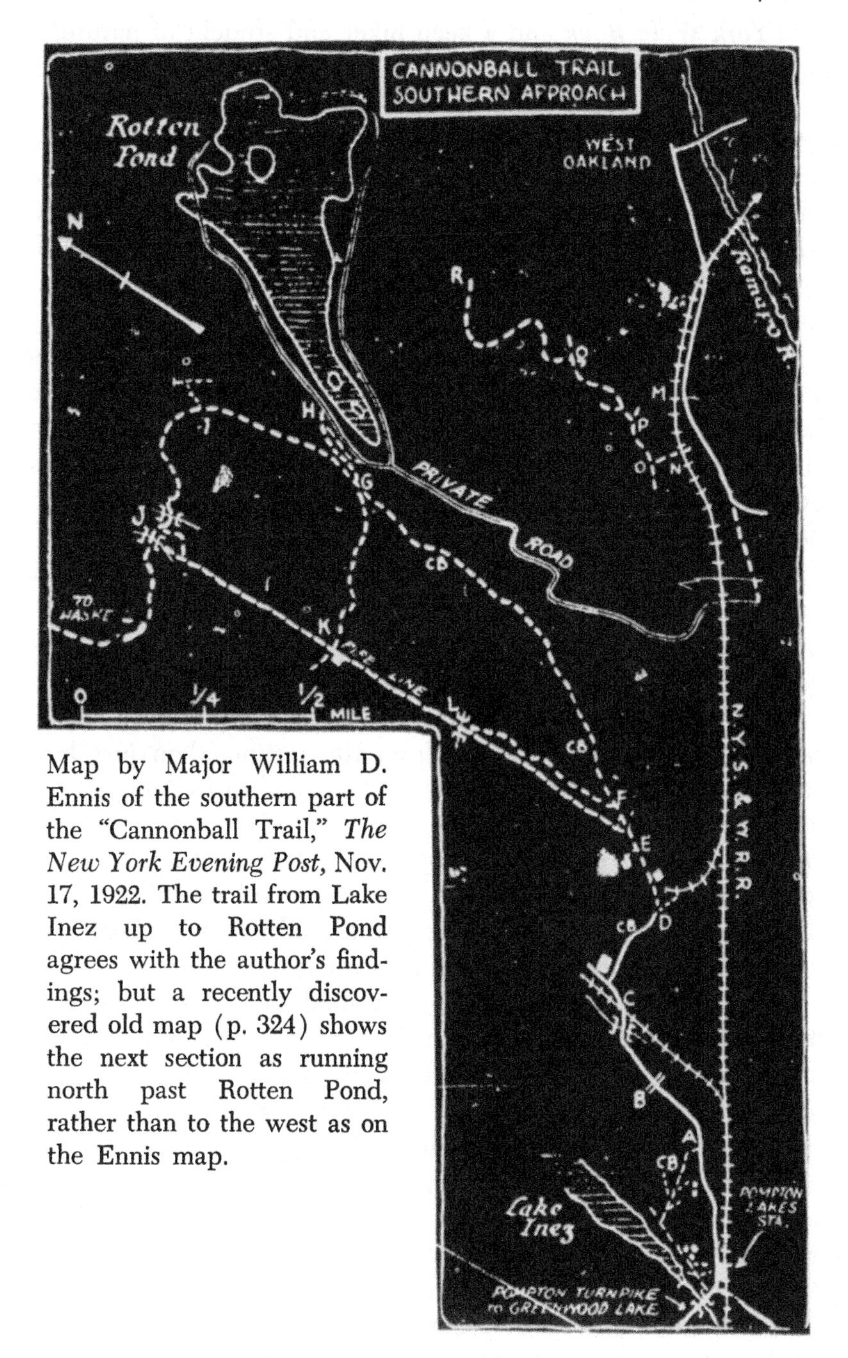

Map by Major William D. Ennis of the southern part of the "Cannonball Trail," *The New York Evening Post,* Nov. 17, 1922. The trail from Lake Inez up to Rotten Pond agrees with the author's findings; but a recently discovered old map (p. 324) shows the next section as running north past Rotten Pond, rather than to the west as on the Ennis map.

New York Walk Book and a keen hiker and student of nature, had explored the various sections of the trail as early as 1913. His contention that the Cannon Ball Road had been a military thoroughfare with distinctively defined features was qualified. "Doubt exists as to its entire route, but parts are identified, especially in Ramapo Township south of Bear Swamp Pond." [4] Place cited that area as evidence for his belief, saying that parts of the roadway had "generally good grade." He seems, however, to have overlooked an extremely plausible alternative for the existence of such a road: generations of mountain people, traveling back and forth, beat endless patterns of trails. With constant use through the years, these narrow trails became clear-cut and widened further by wagon wheels. Yet the exploratory work of Ennis and Place, vague as it often seems, is enough to convince those who choose to believe in the existence of the "secret" road.

Attempts to retrace the road today lead to rather puzzling conclusions. Apparently sound historical evidence can be found for connecting such a road with the Pompton area for a street near the Pompton railroad station is "Cannonball Road." A case can also be made for a somewhat worn trail, or lane, there. To locate it, one goes west about 90 yards from the station, then just before reaching the bridge over the Wanaque River, turns north into the woods on the trail following the east bank of the stream. Approximately 175 yards past the nearby dam, a fork on the hillside overlooks Lake Inez. By taking the right fork and avoiding turn-offs one soon finds the trail descending to a roadbed, worn several feet below the surface of the ground, indicating use over a good period of time. (In 1922 when Ennis described his journey along this stretch of the old trail, he remarked that it "leads you to a backyard where there is probably a wash hanging out." Singularly enough, twenty-eight years later, retracing Ennis's route, the author found the same backyard—and laundry hanging out!)

At one time, this old path evidently crossed the ground where the seemingly ever-present wash was drying. This property is privately owned today, as it was over a quarter of a century ago. Directly in front of the house is the paved street, Cannonball Road. Although this street follows the general direction, it is believed that the dirt path through the woods is more likely the original course of the storied road. Turning north at this point, the modern Cannonball Road leads to the gates of the duPont explosive works. Until 1922, one was permitted to enter the property and thus follow the old trail's path by winding among the buildings, finally exiting through a fence gate on the eastern side. Today, the public is barred entry and one must follow the railroad tracks running east to a point where the wire fence enclosing the powder works turns north. An alternate route can be taken by going east to Perrin Street which dead ends at the railroad tracks exactly at the place noted above.

Following the fence north, one crosses a railroad spur leading into the powder works property; the trail then goes up the hill. Looking through the duPont fence here, just before the locked gate is reached, one can see the old trail on the left. Crossing the open cut, almost opposite this gate, one enters the woods on a trail east; walks a short distance to a fork where one must take the path right, although that to the left looks more promising. The next mile bears definite marks of man-made construction: cuts, fills, and embankments are all obviously an effort to make a roadbed capable of withstanding heavily laden carts and wagons. Ennis called this section "typical Cannonball." Following the old trail a little over two miles from the duPont gateway, one will reach the private paved road at the southern end of Ramapo Lake.[5] Although most of it is uphill, the walk is not strenuous as the old road traverses the wild and uneven terrain by amazingly easy grades, a distinguishing feature of the Cannon Ball route.

Those knowledgeable in the lore generally agree that the part covered so far was the original Cannon Ball Road. The next stage, however, from Ramapo Lake to the Boy Scout camp, Glen Gray,[6] has been more controversial. Some claim that the old road led west from this point down the mountain through the Wanaque-Haskell area, then along the floor of the valley to Midvale, before turning east again and climbing the mountains to Glen Gray. This projected trail ignores the obstacles in the way of heavily laden wagons. Vehicles following this course would have had to ascend almost the topmost ridge of the Ramapos, then turn west and descend, only to climb once again at a point further north. That route would be twice the distance, and much more difficult, than the road on the valley floor from the bridge at Pompton to Wanaque—only about two and a half miles.

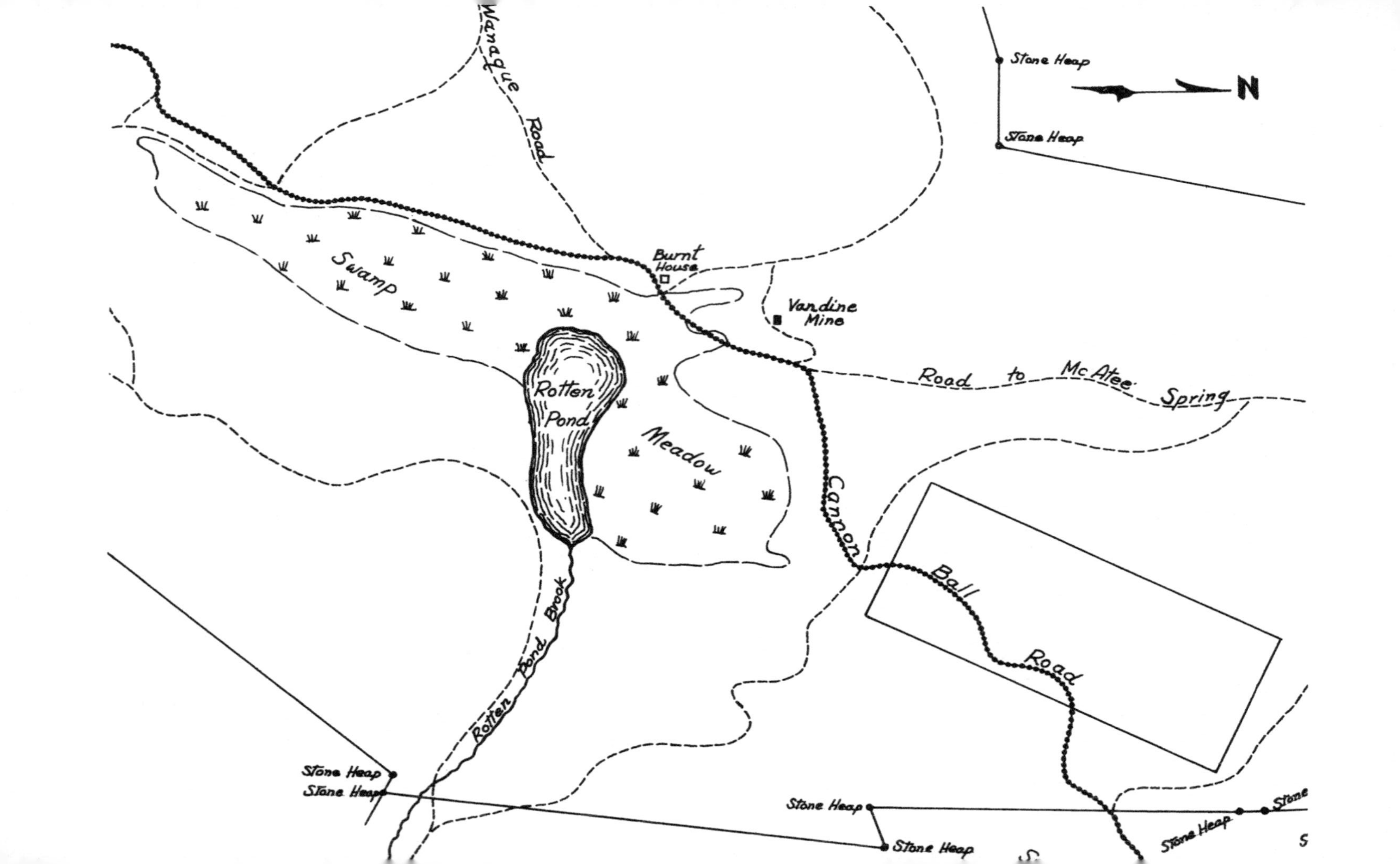
N
Stone Heap
Stone Heap
Wanaque Road
Burnt House
Vandine Mine
Road to McAtee Spring
Swamp
Rotten Pond
Meadow
Cannon Ball Road
Rotten Pond Brook
Stone Heap
Stone Heap
Stone Heap
Stone Heap
Stone Heap
Stone

The recent discovery of the old, undated map mentioned earlier substantiates the author's theory that the Cannon Ball Road continued north through this section closely following the western shoreline of Ramapo Lake.[7] For nearly a half mile the road parallels the private road on the lake side. The map clearly shows that the road did not turn and descend west towards Midvale, as some claimed.

Among other important facts, it reveals Ramapo Lake as a once-swamp meadow where Rotten Pond occupied about a sixth of the present area; and it also shows the Cannon Ball Road cutting across the northwest corner of the swamp meadow and entering the western side of a ravine north of the lake. Almost immediately, the Cannon Ball Road swings east across the bottom of the ravine and starts to ascend the eastern slope, curving north again on its way to the top of the ridge. It soon crosses another private paved road, passing east of Dr. R. F. Decker's home, and slightly northwest of the Arthur Vervaet house.[8] Upon reaching the summit, the Cannon Ball Road runs north just west of a private golf course and parallel to the private road a few yards west. It can be easily followed through the woods and underbrush at this point as it heads toward the property of Valentine Frank. It crosses his front lawn between the house and a clump of trees, continuing north into the woods and eventually crossing the Skyline Drive.

This highway bears a road sign "Entering Passaic County." The private road which also passes the Frank residence enters the highway a few feet south of the sign. Here, across the highway, piles of dirt obscure the continuation of the Cannon Ball Road, but by circumventing the mounds, one soon discovers a distinctly recognizable trail. Here, on the east, the old road parallels the north-bound highway. A state police antenna mast to the left is a good guide to one's position. Despite fallen trees and some undergrowth, the trail is not hard to follow and leads to a clearing where a cross-country pipeline is buried. Entering a wooded section again, the trail alternately goes in and out of forested areas, plunging at last into dense woods, but always remaining on the eastern side of the pipeline clearing. Crossing the pipeline's course, it turns west, then north, continuously crossing and recrossing this landmark until it descends in a northeastern line to Camp Glen Gray. Several other trails leading to buildings and the lake radiate out from the road. The Cannon Ball Road enters from the west, proceeding eastward across the stream and then veers northeast to Camp Yaw Paw and Cannonball Lake.[9]

Some confusion exists in this area where the old road blurs in the maze of trails made by the Boy Scouts. The most obvious trail often leads to a dead end, but one usually finds, through trial and error, the continuation of the Revolutionary road emerging on a well-defined roadbed, for the most part level. The old road passes just west of Camp Yaw Paw northward, soon swings northeast before turning directly east to reach the outlet of Bear Swamp Lake.

The Camp Yaw Paw area was used some years ago for the production of charcoal by a German immigrant, John R. Yahley, who

On the Cannon Ball Road near Camp Yaw Paw.

acquired 172 acres there about 1853. With his wife and seven daughters, Yahley lived in a frame house built close to the Cannon Ball Road and near a spring, said to have never run dry. Only the stone foundation of Yahley's home remains, but the spring is still living up to its reputation. Yahley, who at first planted an orchard and cultivated part of the area, later took up charcoal burning to support his family. He hauled his charcoal downhill in an ox-drawn wagon, using the dirt road which crosses the Cleveland Bridge, spanning the Ramapo River at Valley Road (Route 202). The pasture where his oxen and cattle grazed is now covered by Cannonball Lake. After his death in 1884, the property remained in possession of his widow and children until 1912 when it was sold to a local Boy Scout Council who erected Camp Yaw Paw on the site.[10]

The author was a Boy Scout in the late 1920's at Yaw Paw and recalls with a grin how some of the Tenderfoot Scouts were taken in when they were sent along the old Cannon Ball Road to borrow a left-handed monkey wrench from Camp Glen Gray.

Long a mecca for fishermen, Bear Swamp Lake (a half mile northeast of Cannonball Lake) was visited in April, 1915, by a party that included George E. Carpenter. Having no luck catching fish, Carpenter was strolling around when Charles Conklin, a native guide, shouted to be careful about wandering off and mentioned that he was on the old Cannon Ball Road. This remark greatly intrigued Carpenter and started a life-long interest in the background of the road. In 1947, he wrote of his explorations and research concerned with the subject, and that manuscript is now in the Newark Public Library.[11] Major Ennis, Frank Place, and he were among the first in recent times convinced of the road's existence.

From the dam at the outlet of Bear Swamp Lake, one proceeds along the road following the eastern shore until it reaches an open cut in the woods at the lake's north end. The 1955 edition of the U.S. Geological Survey Map, Ramsey Quadrangle, shows the Cannon Ball Trail here, moving across the open area and into the woods in a northeasterly direction. The trail, as shown on this map, has two possible courses: one ending in the vicinity of the New York-New Jersey State line, east of the Hillburn Reservoir; the other forking to the east as it passes near the north shore of Silver Lake and down into the Hoevenkopf Mountain area at Mahwah. Neither of these seems a logical course for the old Revolutionary trail.

A third course appears the most probable. Starting from the open area north of Bear Swamp Lake, one follows this wide swath in the woods west across the inlet brook flowing into Bear Swamp Lake and soon comes across a trail from the north. This should be followed, disregarding any side trails, to the south end of Cranberry Lake. The latter stretch is a remarkable example of a level, hard solid-bedded road in very rocky mountainous terrain.[12] It could only have been constructed for a definite purpose—probably for transporting supplies, military or otherwise. At this point a wide, clearly defined old wood road winds along the eastern shore of Cranberry Pond passing the summer home of Henry Pierson.[13] A side trail to the east leading off this road at the site of the Pierson home is not a part of the Cannon Ball Road.[14] However, the wood road, a very likely candidate, heads north, swinging around and downhill into the Ramapo Valley where it meets Route 17 at Sterlington.[15]

It is quite possible that the Cannon Ball Road turned south from Sterlington, down the valley for a short distance before crossing the Ramapo River and swinging around Mt. Torne. The logical route would then proceed northeast up Torne Valley, along the west side of Torne Brook, passing east of Pine Meadow and going on to Call Place. From this point it may have continued to King's Ferry on the Hudson River, but which road it took is far from certain. The old road along Torne Brook and through Pine Meadow was used by the Piersons, who built their ironworks at Ramapo in 1795. By using this route and turning off to Ladentown from the meadow area, they had the most direct and shortest way for shipping their iron by wagon to Haverstraw on the Hudson River.

Place felt that a branch of the Cannon Ball Road might have been the road north from Butler with a turn-off to the Hook Road, on to the large flat rocks used by the Indians for threshing, and then crossing to Midvale, where it ascended the western slopes of the Ramapos intersecting the main Cannon Ball Road. Another branch might have started from Ringwood, passing Sheppard Pond on its way to join the main Cannon Ball Road near present Cranberry Pond. This path is shown on one of Erskine's maps, but he does not identify the Cannon Ball Road itself on any of his military surveys.

Still another possibility exists that a branch joined the Continental Road built by army engineers in 1778. This road entered present Tuxedo Park at the south gate, following the eastern shore of Tuxedo Lake, then going across present Wee Wah Lake before continuing to the main road in the Ramapo Valley. Several years ago, when Wee Wah Lake was drained, the remains of this old corduroy road were found practically intact, the old logs still lying as the Revolutionaries had placed them. Claims to having been once a part of the old military trail across the Ramapos might be made by all of these old and more-or-less forgotten roads: the Butler by-pass connecting the main Cannon Ball Road with the Hibernia ironworks; the Ringwood path for possibly carting boom iron for one of the Hudson River chains as well as the equally

An extension of the Cannon Ball Road discovered when Wee-Wah Lake was drained. This old corduroy road is again under water.

important artillery supplies for the Continental Army; and the former Continental Army road furnishing extended cover for the wagons on their arduous trips.

Reconstructing the Cannon Ball Road is like putting together a jigsaw puzzle and offers a similar challenge. At times one wonders whether there are enough pieces; at other times, whether there are too many. It would be so much simpler to have a picture of the original to check—in this case, a map. But one is stirred to seek additional pieces and to rearrange them, hoping all will fit into place and the original form emerge at last.

The reasons for a hidden road's existence are sound, but the documentary evidence weak. There may never have been any more evidence in writing than we have now—very little indeed. The secret has been well kept.

For the present, each quester will have to decide for himself whether the Cannon Ball Road was fact or fiction. But the search will go on by those whose imagination has been fired by the romantic picture of straining oxen pulling heavily loaded wagons up rises, across gullies, and through thickly wooded groves, to help keep the straggling and struggling Patriots in action.[16]

NOTES

I. Forges and Furnaces in Northern New Jersey

Peter Hasenclever and the American Iron Company

1. Peter Hasenclever, *The Remarkable Case of Peter Hasenclever, Merchant*, pp. 3–4.
2. Joseph Tuttle, "The Early History of Morris County, New Jersey," *Proceedings of the New Jersey Historical Society*, Vol. 2, p. 27.
3. Peter Hasenclever, p. 17. Hasenclever always spelled it "Humfray"; others spelled it "Humphrey," "Homfray," or "Humphries."
4. *Ibid.*, p. 18.
5. *Ibid.*, pp. 64–72. This gives the committee's report in full. Also in *New Jersey Archives*, Vol. 28.
6. *Ibid.*, p. 79.
7. Erskine's July 10, 1773 letter to the printer of the *New-York Gazette and the Weekly Mercury* did not appear in the paper until Aug. 9, 1773, indicating that the printer, Hugh Gaine, possibly had reservations about using it.
8. Many years ago, a locomotive on the New York and Greenwood Lake Railroad bore the name "Hasenclever."
9. Gerhard Spieler, "Peter Hasenclever, Industrialist," *Proceedings of the New Jersey Historical Society*, Vol. 59 (Oct. 1941), p. 254.

Ringwood Ironworks

Dates and locations of documents pertaining to property transactions mentioned in this chapter are: Lib. S (Feb. 28, 1740), p. 137, Lib. S (Mar. 11, 1740), p. 140, Lib. S (Apr. 24, 1740), p. 156, at the East Jersey Proprietors, Perth Amboy, N.J.; Lib. L (Feb. 6, 1796), p. 154, Lib. J (Jan. 1, 1798), p. 258, Lib. T (Aug. 28, 1804), p. 125, Lib. Z (Mar. 18, 1807), p. 243, at the Bergen County Hall of Records, Hackensack, N.J.

1. Testimony of James Board given at the proceedings to determine the boundaries of the Wawayanda and Cheesecocks Patents, held at the Yelverton Inn, Chester, N.Y., in 1785. Sanford Durland, present owner of the property, claims it was actually held in the barn there.

2. *Proceedings of the New Jersey Historical Society,* Jan. 1891, p. 126. William Nelson cites a 1737 reference to the "Busseton Forge by Ringwood Cold Spring." This could easily have been Board and Ward's bloomery, located above the state line, because the terms bloomery and forge were used interchangeably. Some deeds and maps indicate Busseton was then on the Sterling property. Ogden tract S–156 refers to it as on a "river that leads from Busselton Forge or Topomopack Pond [Sterling Lake]." It is also called Timothy Ward's forge.
3. Discovered by ten-year-old Robert Tholl of Upper Saddle River, N.J.
4. When Cornelius Board died in 1749, his son Joseph inherited the forge, which was worked until the property transfer.
5. Joseph Tuttle, p. 27. This and the 6½–acre tract are both mentioned.
6. Peter Hasenclever, pp. 5–6.
7. *Ibid.,* p. 80.
8. Joseph Tuttle, pp. 36–7. Ernest Krauss of Waldwick, N.J., an indefatigable researcher, has uncovered much interesting material pertaining to Faesch and Hasenclever. The author hopes that his findings, which have been of considerable help already, will be published soon.
9. Albert Heusser, *The Forgotten General,* pp. 64–6. (Letter dated New York, July 9, 1771.)
10. *New Jersey Archives,* Vol. 28, p. 246.
11. *New-York Gazette and the Weekly Mercury* of March 22, 1773, and the *New-York Journal; or, the General Advertiser* for April 1, 1773.
12. Joseph Tuttle, pp. 33–4.
13. George Clinton, *Public Papers of George Clinton,* Vol. 1, p. 583.
14. *Ibid.,* pp. 643–45.
15. *Ibid.,* pp. 659–61.
16. Erskine established the Bellgrove store, near the New York state line at Mahwah, where Ramapo Road crosses Route 17. The author owns an 1807 deed which mentions the "Bellgrove Lot."
17. A copy of this diary, made in England, was presented to Mr. Alexander Waldron, superintendent of Ringwood Manor State Park, in 1962.
18. E. Martin Ryerson, the fifth and youngest child of Johannis and Cathalyna (Berry) Ryerson, usually wrote his name Martin J. Ryerson, the "J" for his father's name. He was born Nov. 14, 1751.
19. Many of the letters in the Hewitt letter books have faded so badly that they are difficult to read. Cooper Union Institute has typed copies that are easier on the eyes.
20. Walker was at one time President Polk's Secretary of the Treasury.
21. Lack of good transportation had always been a problem at Ringwood. It was to plague Hewitt for the next 20 years.
22. As the furnaces at Ringwood and Long Pond were in ruins, the company planned to use the two they had erected at Phillipsburg in 1847–48.
23. Edward Hewitt, *Ringwood Manor, the Home of the Hewitts,* p. 37. Mrs. Hewitt was Amelia Cooper, daughter of Peter Cooper, president of the Trenton Iron Co., before her marriage.
24. Allan Nevins, *Abram S. Hewitt,* pp. 201–04.
25. *Ibid.,* pp. 205–06.
26. See chapter on Long Pond ironworks for a description of the coordinated activities there.

27. *Directory to the Iron and Steel Works of the United States,* 1878, pp. 17–18. The furnaces are the ones at Long Pond; there were no furnaces at Ringwood at this time.
28. See chapter on the Ringwood Group of Mines.
29. Letter to the author, Nov. 7, 1960. The Pittsburgh Pacific Co. produced iron ore, mainly in Minnesota, but also in Michigan and Wisconsin.
30. This forge site was located by Claire Tholl and the author in the fall of 1961.
31. Edward Ringwood Hewitt, pp. 9, 33. He says, p. 34, that the Manor House was completed by Martin Ryerson in 1810, and that the old Ryerson house ended at the hall. The Hon. Maude Pauncefote, daughter of the British Ambassador to the U.S. (1889–1902), wrote her impressions of Ringwood for the *New York Tribune* in 1905: "The western wing of the present manor is the complete little house built by the Ryersons in 1812." Her article is in William H. and Joseph W. Belcher, *The Belcher Family,* pp. 98–100. Albert Heusser, *The Forgotten General,* p. 153, gives 1807–10 as the erection date.
32. This is the only documented description giving the actual size of the house.
33. Albert Heusser, pp. 153, 193. Edward Hewitt, pp. 33–6, tells how his mother added three buildings—an old store, a chicken coop, and the manager's house—to the mansion and would have added more, but it had reached the road.

Charlotteburg Ironworks

Dates and locations of documents pertaining to property transactions mentioned in this chapter are: Lib. W (Oct. 25, 1765), p. 5, at the East Jersey Proprietors, Perth Amboy, N.J.; Lib. L (Feb. 6, 1796), p. 154, at the Bergen County Hall of Records, Hackensack, N.J.; Lib. E (May 11, 1805), p. 358, Lib. G (Sept. 1, 1807), p. 156, at the Passaic County Courthouse, Paterson, N.J.

1. Charlottenburg is the older spelling and was used by Hasenclever and some of his contemporaries.
2. Macopin Pond is now called Echo Lake. William Nelson, *Indians of New Jersey,* says *macopin,* in Indian, means place where the pumpkins grow. Franklin's Committee spelled it Makapin. Erskine spelled it Makapien on some of his maps. *Dunker* means black or dark. Franklin's group spelled it Dunken.
3. *New-York Gazette,* Sept. 28, 1772.
4. Joseph Tuttle, pp. 26–7. Also in Frank Malone, "Latter Days of Pre-Revolutionary Charlotteburg," *Proceedings of the New Jersey Historical Society,* July 1962.
5. A visit to the furnace site on May 6, 1961, with two friends, Frank Malone and Jim Norman, to try to remove some of the old furnace tie rods embedded in the ground proved unsuccessful. They returned two weeks later to find the spot covered by the reservoir's waters.
6. This was a project by Frank Malone.
7. The society began the excavation of the Middle Forge site in 1960. Jim Norman, Frank Malone, and Ed Lenick have been busy digging. Mead

Stapler has handled the photography, records, and diagrams, as the work progresses. They are to be congratulated on their volunteer efforts.

For two well-detailed maps of the Charlotteburg works and surrounding area, see Erskine's map No. 90a, ca. 1779, and the separate section on Charlotteburg in the lower left-hand corner of Corey's 1861 map of Bergen and Passaic Counties, N.J. See also Hyde's *Atlas of Passaic County.*

Long Pond Ironworks

Dates and locations of documents pertaining to property transactions mentioned in this chapter are: Lib. L (Feb. 6, 1796), p. 154, Lib. J (Jan. 1, 1798), p. 258, Lib. Z (Mar. 18, 1807), p. 243, at the Bergen County Hall of Records, Hackensack, N.J.

1. Called Long Pond in the old days. Hasenclever obtained 5,937.38 acres.
2. On Nov. 1, 1861, the Trenton Iron Co. conveyed the Ringwood and Long Pond works to Edmund R. Miller, Edward Cooper, and Abram S. Hewitt.
3. First reference to the effect that there were two furnaces apparently in operating condition at Long Pond.
4. Edmund R. Miller conveyed his interest in the property and works to Edward Cooper and Abram S. Hewitt on May 1, 1868. The firm became known as Cooper & Hewitt.
5. The order from Long Pond works, dated Aug. 28, 1868, reads: "Yours of 27th received. Three teams were loaded with No. 1 iron going to Trenton before this order for two or three reached us—the balance of the teams will be loaded as directed." The iron made at the furnace was graded from No. 1, which was best, down to No. 6, the poorest. On Aug. 20, 1868, the furnace report showed grades 1, 3, 4, and 5 of new iron totaling about 128 tons on hand, and grades 2, 3, 4, 5, and 6 of old iron amounting to 41 tons.
6. A heat of 350 degrees was then used to make No. 1 iron.
7. People employed at Long Pond in 1869

George Ackerman
Abram Anderson
Arthur Anderson
William Babcock
Daniel Bailey
Stephen Bailey
Townsend Bailey
L. H. Ball
Jacob Banta
Charles Cahill
Dennis Carey
David Caywood
Leonard Cole
R. H. Colfax
William Conklin
Charles Cove
J. L. Cunningham
John Davis
William Davis
Moses Defrase
Charles De graw
Erastus De graw
Henry De graw
Senica De graw
George De groat
John De groat
Moses De groat
William Devon
Elias Dunk
Samuel Dunk
Samuel Dunk, Jr.
Thomas Dunk
John Falone
Nathaniel Finnegan
John Finnegan
John Flemming
William Gould
Alvin Green
Daniel Green
S. E. Gregory
Martin Hand
William Hand
John Harty
Roberty Harty
John Hays
John Heady, Jr.
Thomas Hewitt
Thomas Hewitt, Jr.
Watkin Howell
M. B. Huyler
Ike Kelsey
Mrs. Laird
James Laird
James Laird, Jr.

William H. Laird
E. Lewis
William McGrady
James Margeson
Henry Marshall
John Marshall
Len Mathews
M. Mathews
J. W. May
John Monks
Aaron Montania
James Montross
R. Montross
William Morgan
Aaron Morse
Benjamin Morse
Eben Morse
George Morse
James E. Morse
James W. Morse
John Nixon
Thomas Palmer
William Paterson
Albert Peterson
William Pisce
James Quinn
John Rees
John Ricker
W. Ricker
John Riggot
Charles Rose
A. Ryerson
George Ryerson
Jacob Ryerson
Peter Scofield
Abram Shay
John Sisco
Thomas Sisco
Charles Snyder
George Snyder
Isaac Stagg
John Stagg
A. J. Stalter
Roy Stanback
E. K. Stidworthy
James Storms
Martin Storms
Edward Suffern
George Suffern
Martin Suffern
Peter Suffern
Peter Suffern, Jr.
William Suffern
George Taylor
John Thompson
William Thorp
Joseph Tier
Pat Tully
Isaac Van Horn
Harriman Vannatter
Robert Vannatter
Benjamin White
Samuel White
James Whitehead
D. Whitnour
Thomas Wilson
Nelson Wright
George Zeak

Many of the names above were spelled in several different ways in the ledgers. The author has used the most common spelling in each case.

8. About a month later on Jan. 28, 1870, Hewitt wrote to George: "The accounts show a loss of $12,000 on Ringwood operations last year. As we got a far better price for iron then, than we now can get, I am afraid we will have to stop the works. I confess that I cannot understand it. No estimate that I can make fails to show a profit, and yet at the end of the year there is always a heavy loss. We lost $25,000 in two years."
9. *Directory to the Iron and Steel Works,* 1878, p. 18. Base measurement by author.
10. James M. Swank, comp., *Classified List of Railmills and Blast Furnaces,* p. 18: one furnace at Long Pond was put into blast in Jan. 1873.
11. First discovered by the author in 1956. Rev. Henry C. Beck, in the *Newark Star-Ledger,* Dec. 15, 1957, tells of the trip with the author.

Bloomingdale Furnace

1. The earliest reference to the furnace appears in a survey dated Dec. 8, 1764, made by George Ryerson for John and Uzal Ogden, for a tract of land "lying in the County of Morris about one mile southeast from Bloemendal furnace." *Lib.* A,326, Perth Amboy transcribed surveys, at Passaic County Courthouse.

   Another early reference to the furnace is found in an abstract from John and Uzal Ogden to Ferdinand and Thomas Pennington, dated Nov. 27, 1766, reading in part "with all and singular the furnace and Iron Works at Bloomendale." *Lib.* A,6 Bergen County transcribed mortgages, at Passaic County Courthouse.

A surveyor's report of April 15, 1767, mentioned in William Nelson's *Passaic County Roads*, p. 17, reads: "Road that leads from Bloomingdall Furnace to Charlotteburgh."

2. W. W. Munsell, *History of Morris County*, p. 49. (Spelling given is "Batolf.")
3. The folding map in the *Report of the Commissioners Appointed . . . for the Purposes of Exploring . . . A Canal to Unite the River Delaware . . . With the Passaic . . .* designates the Bloomingdale furnace as ruins. This map is a most important reference tool for anyone researching ironworks near the Morris Canal route. It lists over 70 forges and furnaces, and indicates whether they are operating or in ruins. It covers the area from Morristown on the south to Charlotteburg on the north.
4. In the *Paterson Morning Call*, April 27, 1938, Minnie May Monks wrote that Mr. Isaac Q. Gurnee had shown her the site of the old Ogden furnace in Bloomingdale and that there was only a large mound where the furnace once stood. In Nov. 1962, Ed Lenick of Wayne, N.J., located the mound and discovered part of the furnace buried inside. Ernest Krauss of Waldwick, N.J., whose considerable research on the furnace has been of great help to the author, helped to find its location. In the spring of 1963, excavation was begun here by Jim Norman, Frank Malone, and Ed Lenick, with Mead Stapler doing the charting. At last writing, the lower walls of the boshes, the crucible, and a portion of the outer walls have been uncovered. As the stack itself had fallen years ago, it is impossible to pinpoint its height accurately—a guess would be about 28 feet. It was probably about 27 feet square at the base.

Bloomingdale Forge

1. Photographs and description of the forge's ruins from the *Engineering News*, Dec. 21, 1889.

Clinton Furnace

1. J. Percy Crayon, "Clinton Falls and Furnace," *The Evergreen News*, reprinted in *The Highlander*, Oct. 1959.
2. W. W. Munsell, p. 48.
3. Munsell gives a different reason for closing the works. "While the works were being constructed iron fell one-half or more in price owing to the tariff legislation and Mr. Jackson was obliged to stop operations."

Freedom Furnace and Forge

1. North Jersey District Water Supply Commission, *Report for the Years July 1, 1926 to June 30, 1929*, p. 83. This report also states: "A number of the cast-iron blower pipes were found in and around the furnace, having an inside diameter of 3 to 4 inches, and about 3 feet long." Charles Capen of Green Pond, N.J., made the following measurements before the furnace was dismantled: diameter at top of bosh, 11 feet; height from ground to top of bosh, 10 feet; diameter below bosh, 4 feet; diameter of furnace

opening at top, 3 feet. Lesley gives the inside measurements of the furnace as 12 by 44 feet in his iron guide.

Minnie May Monks, writing in the Paterson Morning Call of Apr. 27, 1938, states: "Peter M. Ryerson also operated two forges about a half mile up the road [from the furnace]—one located near Zeliff Spring on the old Zeliff Farm and the other forge located near the end of Winbeam [Mountain]. At that time a dam extended across the Wanaque River from the end of Winbeam Mountain to Cooper or Sheep Hill, as it was sometimes called. . . ."

2. *Weekly Guardian*, Paterson, N.J., May 20, 1862.

Pompton Ironworks

Dates and locations of documents pertaining to property transactions mentioned in this chapter are: Lib. E (Nov. 11, 1695), p. 233, at the East Jersey Proprietors, Perth Amboy, N.J.; Lib. A (July 8, 1726), p. 32, at the Passaic County Courthouse, Paterson, N.J.; Lib. J (Dec. 15, 1774), p. 402, Lib. J (Apr. 18, 1797), p. 404, Lib. J (Apr. 18, 1797), p. 407, at the Bergen County Hall of Records, Hackensack, N.J.

1. Charles Winfield, *History of the County of Hudson*, p. 532. Also has Schuyler genealogy.
2. Original deed in the possession of the New Jersey Historical Society. Although Schuyler is the only one named, Anthony Brockholst, Nicholas Bayard, and others had an interest in the acquisition and apportioning of the land.
3. W. Clayton and William Nelson, *History of Bergen and Passaic Counties*, pp. 553–54. Because the Indians claimed the whole valley, the land had to be bought from them as well as from the Proprietors to make good the title.
4. John C. Fitzpatrick, ed., *The Writings of George Washington*, Vol. 7, p. 275.
5. *Ibid.*, Vol. 8, p. 5.
6. *Ibid.*, Vol. 19, p. 141.
7. Isaac Leake, *Memoir of the Life and Times of General John Lamb*, p. 241.
8. J. P. Lesley, p. 227. Lesley writes that the rolling mill had 10 furnaces in all, three trains of rolls, three cast-steel fires, and a hammer, and made perhaps 100 tons of steel in 1856.
9. *Ibid.*, p. 35. The furnace was then 12 by 44 feet wide and high inside and was said to be the oldest three-tuyère furnace in the United States. It went out of blast in 1855.
10. *Directory to the Iron and Steel Works*, 1888, pp. 92–3. The steel works consisted of "5 single puddling furnaces, 6 heating furnaces, 42 crucible steel-melting furnaces, 2 trains of rolls, and 5 hammers; water and steam power; 160 pots can be used at each heat in steel works; product, crucible cast steel and railway car springs; annual capacity, 3,000 net tons."
11. William Nelson, p. 421.
12. Author's interview with Mr. Turse, Aug. 29, 1963. His father, John Martin Turse, born in 1840, worked around the ironworks as a young man in the 1850's.

13. Erskine's original map of 1777 is in the Morgan Library and his later one is Map No. 56 B in the New-York Historical Society.
14. This fireback measures 25 by 30¾ inches. Mr. and Mrs. William Gustavson, Franklin Lakes, N.J., have two in their old home. There are others in the immediate Pompton area.

Wawayanda Furnace

1. J. P. Lesley, p. 36. Alanson Haines in his *Hardyston Memorial*, p. 83, speaks of a John H. Brown who had long been associated with the Ames brothers as a superintendent of their ironworks, but does not mention Wawayanda.
2. *Map of Sussex County*, by G. M. Hopkins, Philadelphia, 1860; also *The Sunday New Jersey Herald*, Newton, N.J., Nov. 11, 1962. The latter has a fine article, "Vanishing Village of Wawayanda," by Dick Drew.
3. Peter E. Scovern, "Unspoiled Wilderness," *The Sunday New Jersey Herald*, Newton, N.J., June 20, 1963.

II. Forges and Furnaces in Southeastern New York

Augusta Works

1. MacGrane Coxe, *The Sterling Furnace and the West Point Chain*, p. 49.
2. Sterling ironworks ledger lists the following as employed in 1792:

Joseph Anderson
Joseph Baley
Addam Belcher
Jacob Belcher
Rev. Jas. Benadict
Sarah Benadict
Benjamin Bennett
James Bennett
John O. Bryant
Henry Call
Patrick Cambel
Moses Craft
Samuel Dean
William Doherty
William Fenton
John Green
Joshua Griffing
James Grubb
Obediah Hunt
John Kinner
John Lasher
Stephen Lawrence
James Lowell
Christopher McCoun
Thomas Miller
William Morrison
Seth Newberry
Ephrem Odle
Joseph Patterson
Anthony Pelser
William Pelser
Barnabus Rinesmith
Michael Wamor
Ephraim Wheeler

3. The frontispiece in Elizabeth Oakes Smith's book, *The Salamander*, shows one end wall and the archway of the forge on the bank of the Ramapo River. A portion of it is still standing today, just east of the historical marker on Route 17, several hundred yards north of the Tuxedo railroad station.
4. For excellent material on Solomon Townsend, see "Captain Solomon Townsend and His Times (1746–1811)," an unpublished thesis by Alyce-Louise Falck.

Cedar Ponds Furnace

Dates and locations of documents pertaining to property transactions mentioned in this chapter are: Lib. B (June 1, 1799), p. 344, Lib. C. (Nov. 17, 1799), p. 30, at the Rockland County Courthouse, New City, N.Y.

1. E. M. Ruttenber and L. H. Clark, *History of Orange County,* p. 811. Queen Anne granted this patent on March 25, 1707, to a group who had bought the land from the Indians on Dec. 30, 1702. Hasenclever was also issued a patent on March 25, 1767, for land in the highlands, west of Bear Mountain, which he acquired for mineral rights. Cheescocks (without the third "e") was the spelling used by Charles Clinton, who surveyed the patent, but the spelling Cheesecocks appears more widely accepted.
2. The following were owners or part owners at one time or another after Samuel Brewster's death: George S. Allison, Henry Beebe, James Blackstock, Liman Bradley, Richard W. Brewster, Samuel Brewster of Putnam County, John T. Bulson, Nathaniel Church, Jr., Eli Gurnee, William Knight, Thomas Murphy, and David M. Prall.

Dater's Works

1. John Y. Dater of Ramsey, a descendant of the ironmaster, visited the Dater Mine with the author. It lies less than a mile southeast of the Tuxedo railroad station, high above the Kakiat Trail. A pillar of rock, left standing in the middle of the 28-foot wide flooded opening, supports the mine roof. The entrance faces to the south-southwest and is just off the crest of the hill. Dates of the mine's opening and closing are not known and neither is its depth. This isolated deposit appears on maps of the Palisades Interstate Park region.

Forest of Dean Furnace

Date and location pertaining to the property transaction mentioned in this chapter is: Lib. C (July 30, 1754), p. 142, at the Orange County Courthouse, Goshen, N.Y.

1. From the original letter in the author's possession.
2. In Griffiths' advertisement of 1770, he said the lease expired Sept. 26.
3. *Calendar of Historical Manuscripts Relating to the War of the Revolution,* Vol. I, p. 656.
4. A most interesting statement, but not true. The Sterling ironworks were much larger and better equipped for production than the Forest of Dean furnace. The Government gave the Sterling proprietors the huge order for the West Point chain in 1778. Boyd and Griffiths (and others) may have thought the Sterling works were in New Jersey. The boundary dispute was not settled until 1769 nor the land confirmed to the rightful owners by the General Assembly of New Jersey until Sept. 26, 1772.
5. *Calendar of Historical Manuscripts,* Vol. I, p. 656.
6. George Clinton, Vol. I, pp. 674–75.
7. Letter to the author, Dec. 15, 1961, from Sidney Forman, librarian, United States Military Academy, West Point, N.Y.

Greenwood Furnaces

1. Lib. N, p. 402, at the Orange County Courthouse, Goshen, N.Y. Earlier deeds, also at Goshen, Lib. E, p. 166. Lib. P, p. 431, and Lib. J, p. 109, show that Moses Cunningham was on the scene as early as 1792, buying iron properties northeast of Mombasha Pond, which he resold in 1798 and 1805, and then disappeared from the picture.
2. Daniel Freeland, *Chronicles of Monroe in the Olden Time,* p. 74. In these early days the Greenwood ironworks was often referred to as the Orange Factory.
3. *Lib.* FF, p. 383, Orange County Courthouse. Gouverneur and William Kemble purchased the tract from Richard Trimble of Newburgh, N.Y., who undoubtedly acquired it from Nathaniel Sands. The Kembles paid Trimble $8,000 for it. (Both Kreutzberg and Parrott err on the purchase date.)
4. *Lib.* 76, p. 476, Orange County Courthouse. The Kembles each had one-third interest, as did Fenwick. Robert Parrott's brother, Peter Pearse Parrott (1811–96), became associated with Robert and the Kembles in 1837. He married Mary Antoinette Arden, after whom Greenwood was later named.
5. J. P. Lesley, p. 5. Base measurement by author. See also E. C. Kreutzberg's fine article, "Orange County Iron Making," *Iron Trade Review,* July 31, 1924, pp. 285–88, for a detailed account of the Greenwood ironworks.
6. Richard D. A. Parrott, *Cold Blast Charcoal Pig Iron,* pp. 14–15.
7. The ruins of the kiln, stamping mill, sawmill, coal house, dam, and raceway can still be seen along the brook, a half mile east of the old furnace ruins.
8. Richard Parrott, pp. 7–11.
9. Copy of letter from the Parrott Iron Co., Oct. 15, 1881, in the author's possession.

Monroe Works

1. Daniel Freeland, p. 34. The locality was called Cheesecocks from 1764 to 1801; then Southfield from 1801 to 1808; then Monroe, after James Monroe; and in 1864, it became Southfield again.
2. Advertisement, *New York Evening Post,* Sept. 13, 1810.
3. The end of the advertisement noted that "castings of every kind would be made on short notice at the Union Air Furnace at the upper end of Broadway."
4. Professor Beck was employed by New York State to inspect the mineral districts of Orange and Rockland Counties.

Noble Furnace

1. The furnace is approximately 21 feet square at the base and close to 25 feet in height. It has two arches, one of which has crumbled. The other arch, which faces the stream, measures six feet in width at the front opening and narrows down to a width of less than four feet at the interior end.

Queensboro Furnace

1. Perhaps because the Forest of Dean and Queensboro furnaces were often both called the Orange furnace or because they were already under the same ownership by this time: the former, however, was not in operation then and had been out of blast for a number of years. See chapter on the Forest of Dean furnace.
2. *Outing and the Wheelman*, Vol. 5, p. 170; a good picture of the furnace, which was 25 feet square at the base, is on p. 165.
3. W. A. Rigby, address given at Palisades Interstate Park Conference, Lake Sebago, N.Y., May 3–5, 1940.

Ramapo Works

1. Edward Pierson, *The Ramapo Pass*, pp. 102–03. Also letter dated April 15, 1881, from Henry L. Pierson to James Swank, copy owned by author.
2. Pierson Mapes, who lives in the valley near the site of the old works, thinks the Piersons built the road up the Torne Valley.
3. See Eben Cobb's fine chapter in Cole's *History of Rockland County* for additional data, and a diagram of the Ramapo works.

Southfield Ironworks

1. MacGrane Coxe, p. 51.
2. Southfield furnace ore receipt book lists as drivers of ore carts: Peter Conklin, George Patterson, Robert Ketchum, John Davie, Joseph Wimar, and Jeremiah Conkling.
3. Raphael Hoyle (1804–36), who showed promise of becoming one of the nation's finest painters, died a year after completing this painting owned by author.
4. Southfield furnace blast account ledger gives these statistics for Blast No. 1 in 1859:

Dimension of Hearth Square	29 in.	No. of Charges in Casting	25
Height of Tweers & Dam	16 in.	No. of Charges in 24 hours	25
Length of Hearth	4½ ft.	Bushels of Coal to Charge	22
Diameter of Boshes	12½ ft.	Pounds of Flux to Charge	35
Angle of Boshes	60 deg.	Augusta Ore used	490
Time of Casting	9 PM	Product of Casting	1 ton, 15c

Sterling Ironworks

Dates and locations of documents pertaining to property transactions mentioned in this chapter are: Lib. E-2 (Oct. 1, 1736), p. 124, Lib. S-1 (Nov. 10, 1736), p. 118, Lib. S-1 (Feb. 13, 1739), p. 134 at the East Jersey Proprietors, Perth Amboy, N.J.; Lib. 178 (Apr. 1, 1864), p. 196, at the Orange County Courthouse, Goshen, N.Y.

1. Also spelled "Topomopack" and "Tomopack" in early surveys.
2. See Ringwood Ironworks chapter, note #20.

3. Testimony given in the New York-New Jersey boundary-line dispute in 1769:

   ISAAC VAN DEURSEN, age 71 years. That he is acquainted with Sterling Iron Works and has known same about 20 years or more. That Timothy Ward and Nicholas Colten first owned and built the same and that they were built about 20 years or more, but how much longer he cannot tell.

   DAVID HENNION, age 39 years. Knows Sterling, Ringwood, and Long Pond Iron Works and are supposed to be in New Jersey government. That at Sterling Iron Works there are 1 furnace, 1 forge, 1 saw mill, and 1 grist mill erected. That at Ringwood there are 2 forges, 1 furnace, and 1 saw mill erected and that at Long Pond there are 1 furnace and 1 forge erected. That Ringwood and Sterling iron works have been built as long as he can remember, but Long Pond Iron works have not been built longer than 3 years.

   DENNIS MOORE, age 42. That he knows the tract of land belonging to Messrs. Hassenclever and Company at Ringwood and Long Pond. That he was present at a survey of it and that it contained as he has been informed 10,600 acres. That he has been informed that there were between 300 and 400 people employed on said tract at one time; that some of them paid taxes and other duties to New Jersey; that many of them were single men; and that some of said persons who are workmen at said iron works live in houses erected for them by proprietors of said works.

   Howard Durie located the above information for the author. Budke Vol. BC 29, Manuscript Room at the New York Public Library.
4. Samuel Allinson, comp., *Acts of the General Assembly*, p. 370.
5. In 1838, Peter Townsend told Dr. William Horton that Abel Noble built the forge. If he was referring to Abel, son of William Noble, it was unlikely for Abel was only 15. No doubt it was his father, who later went into partnership at Sterling with Hawxhurst.
6. *New-York Mercury*, Dec. 24, 1759.
7. *Calendar of Historical Manuscripts*, Vol. I, pp. 446–47.
8. *Ibid.*, Vol. II, p. 69.
9. See chapter "Chains Across the Hudson."
10. Matthew Davis, ed., *Memoirs of Aaron Burr*, Vol. I, p. 173.
11. Since Peter Townsend, Sr., died in 1783 and as his son was only 19 or 20 in 1789, the time of Crèvecoeur's supposed visit, he could not have had the reputation described. This and other incidents Crèvecoeur relates make one wonder whether he actually visited the area at this time, or whether he was remembering an earlier trip. Peter, Sr., willed the ironworks to his wife, Hannah Hawxhurst, and under her management and that of her sons, Peter and Isaac, all went well.
12. An overestimation: it is closer to 310 acres. It was unintentional, because he wrote it was a "little lake."
13. This is long for the length of the bellows.
14. Crèvecoeur, Vol. I, pp. 282–87. Sections pertaining to the Ramapo area have been translated in a booklet, "Eighteenth Century Journey Through Orange County," pp. 29–31, as well as in MacGrane Coxe's *The Sterling Furnace and the West Point Chain*.

15. MacGrane Coxe, opposite p. 16.
16. Employed at Sterling Furnace between June 1792 and May 1794:

Ackerman, Abraham	Cornel, Tobias	Morgan, Thomas
Anderson, Joseph	Cornelious, David	Mushsell, Cornelius
Anthony, George	Crips, Peter	Odle, John
Anthony, John	Davey, John	Oldrage, Christopher
Badcock, David	Davey, Thomas	Oldrich, James
Badcock, Isaac	Dearmond, George	Osborne, Elizah
Badcock, John	Degraw, Harman	Patterson, John
Bains, Cornelius	Degraw, Thomas	Robash, Daniel
Ball, David	Demorest, John	Roberts, Davey
Ball, John	Dikins, David	Russell, Wm.
Barns, Cornelius	Easton, Henry	Silvester, Isaac
Bearmore, Adam	Everman, John	Slippy, David
Bearmore, Henry	Finton, Solomon	Smith, John
Bearmore, John	Finton, Wm.	Smith, Robin
Benjamin, Jacob	Fish, Robert	Smith, Wm.
Bennett, Levi	Fitzgerald, Wm.	Springsteel, John
Bennett, Nat	Garrison, David	Star, Aaron
Brooks, David	Garrison, Wm.	Stephens, John
Brooks, George	Green, John	Strickland, Caleb
Brooks, Michael	Green, Peter	Taylor, Edward
Burk, Bartholomew	Hallock, Jonathan	Tebow, Andrew
Burrows, Thomas	Hartly, James	Tebow, Peter
Butler, James	Hollister, Wm.	Townsend, Zebulon
Byron, Thomas	Humble, John	Trickey, Christopher
Cachall, Thomas	Jennings, Redmond	Trickey, Wm.
Camble, Patrick	Johnson, Titus	Utterman, James
Caraiss, Thomas	Johnson, Wm.	Wagon, John
Clay, Christopher	Kilsy, Abraham	Wallis, John
Conkling, Elias	Kincaid, Robert	Weymon, Michael
Conkling, John	Kirkindall, John	Williams, Ben
Conkling, Sam	Mattey, Jesse	Williams, John
Cooney, John	McMurtory, Alexander	Williams, Joseph
Cooper, James	Morgan, Daniel	Willson, James

There were many variations of the spelling of personal names. In some cases, certain names in the same ledger were spelled in as many as four different ways. The one used most is given here.

Employed under Abel Nobel at Sterling during the same period:

Cooper, Joseph	Jensen, Wm.	Priestly, Thos.
Delancy, Abraham	Jonson, Ben	Smith, Solomon
Dell, George	Langly, Sam	Spring, Stephen
Duffer, Caty	McMichl, Clemons	Stuperfell, Henry
Finch, Solomon	More, Wm.	Wagon, Michael
Gordon, David	Morris, Wm.	Writingour, Peter
Jensen, Peter		

17. The New York State chapter of the Daughters of the Revolution graciously erected the memorial and it is hoped that someday they can take steps to

correct the error. At present the plaque is lying on the ground about 50 feet or more east of the furnace, which was restored in 1958. Moved aside when the rebuilding of the furnace was started, it was never put back.

18. It was listed in the Cadmus Bookshop (New York, N.Y.) catalogue No. 96, item 263 (1930).
19. The earliest date substantiated by available records is 1736.
20. In 1882, when a post office was established, the name was changed to Sterlington. James Everett Ward was station agent at Sterlington for 54 years, retiring in 1940 when the station was permanently closed.
21. Frank Green, *The History of Rockland County*, pp. 397–98. In 1882, the gauge was altered to 4 ft. 8½ inches. The road opened Nov. 1, 1865.
22. Because the property covered such a large area, the company had to pay taxes in the towns of Warwick, Ramapo, Tuxedo, and Monroe, N.Y., as well as in the Township of West Milford, N.J.
23. Told to the author by Tom Whitmore and Charlie Edwards, the caretakers at Sterling, who know a great deal about the works. Mr. Edwards died recently. Tom Whitmore is still there.
24. *New York World-Telegram and The Sun*, Apr. 13, 1965.

John Suffern and His Ironworks

1. Document owned by author. Evidently John Suffern began erecting the forge before he made the agreement with his son Andrew, because Jeremiah H. Pierson of the Ramapo Works, writing about the proposed Nyack Turnpike on Feb. 19, 1806, says: "I find from an examination of our books that we are transporting to and from the landing from our works only 872 tons, on an average, in one year; a saving of five miles in the distance will therefore save the transporting of one ton 4360 miles. To which add the freight of all Sterling Works, Ringwood, Dater's and the forge building by Mr. Suffern, and the road is of considerable magnitude to this part of the community."
2. The forge at New Antrim was located on the west bank of the Mahwah, south of the old Nyack Turnpike, according to David Cole's *History of Rockland County*, p. 278; Frank Green's *History of Rockland County*, p. 395; and Arthur Tompkins's *Historical Record of Rockland County*, p. 541. However, David H. Burr's 1829 *Map of the Counties of Orange and Rockland* shows a forge site on the east bank. The 1839 edition of the map labels the forge "Suffrees" and places it on the east bank just north of the Nyack Turnpike, with the route of the newly constructed New York & Erie Railroad almost passing over the forge site. Suffern also had a gristmill, sawmill, and woolen mill along the stream.

Woodbury Furnace

1. E. C. Kreutzberg, "Orange County Iron Making," *Iron Trade Review*, July 31, 1924, p. 285. Richard D. A. Parrott, in his pamphlet, *Cold Blast Charcoal Pig Iron*, gives 1830 as the date of the furnace's erection.

III. The Mines

The Miners

1. Mr. Green was interviewed by C. Boehm Rosa of the *Middletown Times Herald,* who wrote an interesting series of well-illustrated articles on mining in Orange County, in April and May, 1938.

Greenwood Group of Mines

1. Raphael Pumpelly in *Report on the Mining Industries of the United States,* p. 93. Colony doubts the mine was this deep.
2. Heinrich Ries, *Report on the Geology of Orange County,* p. 452. (Ore from the mine was used at Southfield and Greenwood furnaces, either alone or mixed with O'Neill and Long Mine ore, and made iron of excellent quality, according to Lewis C. Beck.)
3. From notes taken in July, 1950, by G. H. Bobeniether of Hawthorne, N.J.
4. Charles Clinton, surveying this area in 1737, wrote in his field book ("The Marble Book") that Indians had discovered iron nearby.
5. Heinrich Ries, in his report on Orange County, wrote that the O'Neill Mine had not been worked for 16 years. Charles Green (interview with C. Boehm Rosa) thought that it closed in 1885 or later.
6. If this is the same mine that Lewis C. Beck refers to in his diaries as the Rich Mine.

Hudson Highlands Group of Mines

1. According to Samuel Eager, p. 596, the ore from the mine went into part of the *chevaux-de-frise* at "Pallopel's Island," extending toward Plum Point at the mouth of Murderer's Creek.
2. *Port Jervis Gazette,* July 31, 1893.
3. With the author were Eben Gould, Clair and Richard Sawtelle, and Ada Mead, now the author's wife.
4. Christie made the survey for the owner, A. Lawrence Edmands of Newton, Mass., in 1891. Copy owned by Harold Sherwood, a friend who is a Spring Valley, N.Y., attorney and specialist on land titles. Mr. Sherwood recalls that Edmands was associated at the time with Thomas Edison in iron interests.
5. See Cedar Ponds furnace, note 2.
6. Frank Green, p. 162, writes: "The works erected because of this mine consisted of a furnace a short distance above the present lowland bridge, a foundry on Florus Falls Creek, a half mile further west, a forge or a bloomery on the property now owned by Henry Goetschius, a bloomery a short distance from the old Slutton House, and still another bloomery and furnace just below the outlet of Cedar Ponds."

Ringwood Area Mines

1. Date given author by Lovell Lawrence, then manager of the Ringwood Co., in 1939.
2. Told to the author and his wife at the mine site.
3. A 26-inch snowfall, 1947–48, submerged the operation, proving too much of a handicap.
4. Col. Sanders believed that he had found more iron ore in the earth in the Peters Mine than had been removed in the mine's history.
5. This notice is now in the author's collection.
6. Since writing the chapter on the Ringwood mines, another change in ownership has taken place. In October of 1964, the Pittsburgh and Pacific Co. sold the 900-acre tract to the Ringwood Realty Co., a division of the J. I. Kislak Co. of Newark. They plan to seal off the mine entrances and build a $50 million self-contained community on the tract.

Sterling Group of Mines

1. Heinrich Ries, p. 472; book contains a picture of this mine, and of others.
2. Interview with Tom Whitmore, watchman at Sterling, who is full of information on the area.
3. Interview with Col. Lewis Sanders, mining engineer, in 1962: his experience with the mines at Sterling and Ringwood goes back 50 or more years.
4. Ledger in author's collection.
5. William Horton, *Report of W. Horton to W. W. Mather on the Geology of Orange County*, pp. 163–64.
6. *Ibid.*, p. 173.
7. The booklet is in the library of the American Iron and Steel Institute, New York, N.Y.
8. William Horton, pp. 172–73.
9. Not the brook that runs from the outlet of Sterling Lake, but a small stream in between the mine entrance and the outlet brook at that time.

Whynockie Area Mines

1. Author's interview, 1939, with Mr. Lanning, who lived in the Whynockie Valley. The author prefers the old name to the present name, Wanaque, a corruption of Whynockie. Martin Ryerson used the spelling Whynockie when referring to his ironworks or the area. Some others spelled it Wynockie without the "h."

IV. Chains Across the Hudson

1. E. M. Ruttenber, *Obstructions to the Navigation of Hudson's River*, pp.7–8. Punctuation and capitalization have been modernized.
2. *Ibid.*, p. 8. The members included were: Col. Robert Hoffman, Henry Glen, Alexander McDougall, and William Paulding.

3. *Ibid.*, p. 13.
4. On Aug. 16, 1776, fire ships covered with combustible materials (melted pitch, straw, turpentine) were tried out against the English frigates, *Rose* and *Phoenix,* in the Hudson opposite Yonkers. After dark they moved up silently with the tide and approached the enemy ships. Shortly after the sentinels on board had announced "all is well," two of the fire ships burst into flame, one at the side of the *Phoenix,* one against the other vessel. Some damage was inflicted and Washington stated two days later: "Though this Enterprise did not succeed to our Wishes, I incline to think it alarmed the Enemy greatly. . . ."
5. E. M. Ruttenber, pp. 65–6, 68–9. In their early meetings and letters, the committee used the term "boom" even though they meant "chain." At their July 25 meeting a discussion took place as to the technical differences between the various types of barriers, in which it was resolved that "boom" and "chain" were two different things. In view of this, instead of a "boom," the artisans were directed to make and "fix a chain" across the river. It is said that the iron furnished by Robert Livingston for the chain came to about 22 tons.
6. *Writings of Washington,* Vol. 5, p. 15.
7. E. M. Ruttenber, pp. 68–9.
8. *Ibid.*, p. 76.
9. Peter Force, ed., *American Archives,* Vol. I, p. 78.
10. *Writings of Washington,* Vol. 7, pp. 444–45.
11. *Ibid.*, Vol. 8, pp. 114–16.
12. Robert Beatson, *Naval and Military Memoirs of Great Britain,* Vol. 4, p. 236.
13. Erskine's cash book shows that in July, 1777, he received from General Mifflin £2,960 for 37 tons of bar iron delivered to Peekskill; in Sept., 1777, he received on account boom iron, £1,300 and in May, 1778, the entry reads, "To Quartermaster General, in full for Boom Iron Account £3860.17.2½."
14. Iron from Col. Robert Livingston's blast furnace at Ancram Creek and the Sorel river chain made up the Fort Montgomery chain. The total sum for the boom iron made at Sterling was just over £5,945. Installments totalling £5,747 of this amount were paid, but the full amount never collected.
15. *Journal New York Provincial Congress,* Vol. I, p. 1117.
16. The contract specified:

    1. Delivery before April 1, 1778, or sooner if possible, an iron chain.
    2. A total length of five hundred yards, each link two feet long and a cross-section of two and one-quarter inches square or as near to this size as possible.
    3. The chain is to contain a swivel every 100 feet and a clevis every 1000 weight.
    4. To deliver at least twelve tons of anchors of such size as directed by Hugh Hughes or successor, the material to be of the best Sterling Iron.
    5. The United States shall pay 440 pounds per ton if produced within the contract date. If a longer time is required the price is to be 400 pounds per ton. Payment is to be made in such proportions as the work is delivered.
    6. A competent committee composed of three to five men, uninterested

in the company, is to judge the quality of the iron and if condemned, the company must make good.

7. If mechanical failures occur, the company must make good.
8. The company is exempted from all military duties for nine months.
9. Seven fires at forging and ten at welding are to be kept in operation.
10. Labor is to be supplied by the army if other labor cannot be secured. However, army labor wages must be deducted from the contract figure.

17. This could not be true, since the Ringwood iron arrived after the Fort Montgomery chain and boom were placed in the river. He certainly does not speak well of his neighbor's iron. Robert Erskine's Ringwood works lay just a few miles south of Sterling. In the author's opinion, it is further proof that if the iron was of poor quality, it was not Ringwood iron that was used: Ringwood iron had a good reputation.
18. Actually, Peter Townsend was only part owner and it was Hugh Hughes, not Col. Pickering, who arrived to sign the contract for the chain. Lossing and other writers err here. The contract itself was signed by Hughes and Townsend, in the presence of the witness, P. Tillingham.

    He is an entire month off: the contract was signed on Feb. 2, 1778. Furthermore, it was Monday, not Saturday. This statement must have misled Lossing and others.
19. E. M. Ruttenber, pp. 135–36.
20. *Ibid.*, pp. 142–43.
21. Frederick C. Haacker of Mount Vernon, N.Y., thought it probably took 10 to 11 weeks. Mr. Haacker's notes on obstructions in the Hudson River during the Revolution, now in the New York Historical Society, represent a lifetime's work and are most important source materials. In view of the excellent research job, it is to be regretted that Mr. Haacker did not publish a book on his findings. At least his material is available for future researchers. Peter Townsend II told Dr. Horton it was made and delivered in six weeks. William Horton, p. 174.
22. E. M. Ruttenber, pp. 139–41.
23. Carl Van Doren, *Secret History of the American Revolution,* pp. 460–61.
24. Rufus Harte calls him John C. Abbey, but Abram Hewitt's correspondence mentions Westminster Abbey.
25. When Hewitt acquired the Ringwood property for Peter Cooper in 1853, he noticed two old rusty links, the same Martin Ryerson had found there when he bought the ironworks in 1807. They were probably leftovers from Revolutionary days. In 1876, Hewitt loaned them to the Centennial Exposition in Philadelphia. Unfortunately, they were not returned and have never been found.
26. It is astonishing that Hewitt, who ran the ironworks at Ringwood for a long time, did not know that the contract for the West Point chain was not made until Feb. 2, 1778, and the chain itself was not set in place across the river until the latter part of April.
27. Roscoe Smith of Monroe, N.Y., later purchased the chain links from the Coxe family and they are now exhibited at his "Monroe Village"—well worth a visit.
28. There were only two chains across the Hudson River during the Revolutionary War. Writers who say three are wrong and are either handing

down an error of early days, or have confused the chains with different types of obstructions, such as *chevaux-de-frise*, etc.

29. Edward Ringwood Hewitt, Abram's son, in his *Ringwood Manor, The Home of the Hewitts*, p. 46, writes this of the chain: "It was secured from a junk dealer named Westminster Abbey, who guaranteed it. . . . In later years it was recognized by an English iron manufacturer as one of the Admiralty Buoy chains which had been placed in New York Harbor, and had been made by his firm. I analyzed the iron of the links and found it was not Ringwood iron, but Lowmoor iron from England. This together with the fact that the links are not of the same length as those of the original chain at West Point, seems to establish the fact that this is not part of the West Point chain. . . ."

V. The Cannon Ball Road

1. William Nelson, *History of the City of Paterson and the County of Passaic*, p. 421, writes of the ironworks at Pompton and the cannon balls made there for the American army. "These munitions of war were carted by a circuitous route through the Ramapo Valley or Paramus and hence locally known to this day as the 'Cannon Ball Road.' . . ." A footnote adds: "A very difficult old wood-road leads on top of the mountain directly north of the Pompton Lakes railroad station northerly to Ringwood, and local tradition says that this was used during the Revolution by the Ryersons, great ironmasters, for the secret transportation of cannon balls to the British. In answer to this, it may be noted that while this section of five or six miles of road might serve for secret transportation, there would remain twenty miles or more of open country through which to cart the cannon balls. Secondly, the Ryersons did not own the iron mines during the Revolution. Thirdly, the story had its origin during the War of 1812. In 1821, Jacob M. Ryerson, who then owned the mining property, traced the report to two well-known citizens and compelled them to acknowledge in a Newark newspaper of the day over their own signatures that they did not believe there was any basis for the rumor."
2. A few years ago, Fred Reinhardt noticed a strange spherical object imbedded in the ground in the vicinity of the old Cannon Ball Road between the duPont powder plant and Rotten Pond. Prying it up, he found it to be an old cannon ball. Though we do not know why the cannon ball was there, judging from Reinhardt's discovery, the road's name seems to have some basis in fact.
3. In particular, see the issue of Nov. 17, 1922.
4. Raymond Torrey, Frank Place, Jr., and Robert L. Dickinson, *The New York Walk Book*, p. 259. Frank Place has some interesting remarks on the old road in *The New York Evening Post*, Nov. 17, 1922.
5. Formerly called Rattan Pond before being corrupted to Rotten Pond.
6. Boy Scout camp of the Eagle Rock Council which represents the Montclair, Glen Ridge, Verona, and Caldwell districts.
7. R. F. Decker, who lives on the ridge at the north end of the lake, showed the map (a copy of an older one) to the author.

8. There is a beautiful view from the home of Arthur Vervaet, as well as from the Deckers', where one can look down on sparkling Ramapo Lake and miles of surrounding country. This area is all private property; and through Mr. Vervaet's kindness, the author was able to explore the region on many different occasions.
9. Scout camp for boys from the Ridgewood and Glen Rock, N.J., Council.
10. They made the pond and named it Cannon Ball Lake. The data on Yahley is from an article by Elliot Hempstead in the *Bergen Evening Record,* Sept. 19, 1953.
11. Unfortunately, some information given to Carpenter, and which he used was misleading. Another paper on the Cannon Ball Road, by the late John Storms, and also in the library, has some errors.
12. Mr. Seely Ward of Chester, N.Y., who lived in the Sterlington area during his boyhood, showed the author the route of the Cannon Ball Road from the north end of Bear Swamp to the south end of Cranberry Pond. This section, which has always been a puzzle to historians, was explored in the spring of 1961.
13. Mr. Pierson told the author how his father had always spoken of the old road along the eastern shoreline as the Cannon Ball Road.
14. According to Pierson Mapes, the side trail here was built by Piersons, who first came into the area in 1795 and established the Ramapo ironworks.
15. Frank Place believed the road ran from Bear Swamp north to the valley south of present Potake Pond, then around the west side of High Mountain and on into Sterlington, and did not go into further detail, but Place's road is much closer to the route pointed out by the author than the path shown on the Ramsey Quadrangle map of the U.S. Geological Survey.
16. Any exploration of the Cannon Ball Road should be done with the aid of a map as it is very easy to get lost. Watch out for snakes in summer.

GLOSSARY

Adit—A nearly horizontal passage from the surface through which a mine is entered and dewatered. In this country an adit is usually called a tunnel.

Anchory—A forge where anchors are made.

Blast Furnace—Usually a stone stack shaped like a truncated pyramid where ore is smelted; from 20 to 30 feet square at the base and 25–55 feet in height. Sometimes called a stack or shaft furnace.

Blast Pipe—A pipe for supplying air to the blast furnace.

Bloom—1) A semi-molten lump or ball of pasty crude iron. 2) A rough bar of iron drawn from a bloomery ball for further manufacture. 3) A mass of iron or steel formed by consolidating scrap at a high temperature by hammering or rolling.

Bosh—The widest part of the inner chamber of the furnace.

Bridge—A trestle or platform over which fuel and ore are conveyed to the mouth of the furnace stack. It usually ran from the side of a hill to the top of the stack.

Buckstay—An upright steel or iron brace against the exterior of the furnace for reinforcing the stone and brickwork. See **Tie Rod.**

Casting House—A building at one side of the furnace. It covers the archway through which the molten iron is tapped, and the pig beds and casting sands.

Charcoal Pit—Not actually a hole, but dry, level ground where wood is usually stacked upright in a conical shape, then covered with earth and sod, and set on fire.

Charge—The materials such as charcoal, limestone flux, and iron ore, introduced at one time, or in one round, into the furnace.

Cinder Notch—See **Slag Hole.**

Collier—A worker who makes charcoal. Also spelled coler and coaler in old ledgers.

Crucible—The lower part of a blast furnace used for collecting molten iron.

Drift—A horizontal passage underground following the vein of the ore.

Forge—A small furnace for making wrought iron by heating and hammering the iron mass into desired shape.

Gudgeon—An iron pin used to fasten stone blocks together.

Hearth—The floor or sole of a furnace.

Iron Runner—The channel or trough for conducting molten iron from the tap hole of a blast furnace to the molds.

Lintel—A heavy iron beam for supporting the load of a furnace arch.

Merchant Iron—Iron in the common bar form, convenient for market.

Mill—Any establishment for reducing ores by other means than smelting. More strictly, a place or a machine where ore or rock is crushed.

Millrace—The current of water that drives a mill wheel, or the channel where it flows from the dam to the mill.

Mucker—One who loads mine cars and skips and sometimes acts as a trammer, pushing them to the shaft, tunnel, or adit mouth.

Outcrop—A bed or layer of rock or ore which has come to the surface of the ground.

Pig—A hardened, oblong mass of iron, roughly shaped like a loaf of French bread. See **Pig Iron.**

Pig Beds—The sand bed where furrows are made as molds.

Pig Iron—Crude cast iron from the blast furnace. When the furnace is tapped the molten iron flows down a channel (the sow) molded in the sand, from which it enters lateral troughs—the pig molds. In each bed, the ingots lie against the sow like suckling pigs; hence the name pig iron.

Pillar—A column of ground or mass of ore left to support the roof or hanging wall in a mine.

Puddle—To subject cast iron to heat and stirring in the presence of oxidizing substances for conversion into wrought iron.

Puddle-ball—The lump of pasty wrought iron taken from the puddling furnace to be hammered or rolled.

Reverberatory Furnace—A furnace in which ore is submitted to the action of the flame without contact with the fuel. The flame is reverberated downward from the roof of the furnace, upon the charge.

Roaster—A furnace for roasting the ores until the sulphurous fumes disappear.

Rolling Mill—A place where metal is made into sheets, bars, rails, or rods by working it between pairs of rollers.

Salamander—A mass of fused and partly reduced iron ore in a furnace.

Shaft—A narrow deep excavation made for locating and/or mining ore; raising water, rock, or ore; hoisting and lowering men and materials; or ventilating underground workings. An inclined shaft follows the dip of a vein. A vertical shaft is sunk directly downward.

Shingle—To drive out impurities from the puddled iron by heavy blows from a tilt hammer or trip-hammer.

Skip—An open iron 4-wheeled car which runs on rails and is used especially in inclined shafts. Also applies to a large hoisting bucket. Both are principally used to haul iron from mines.

Slag—The molten non-metallic layer formed by the reaction of the flux and the gangue floating on the surface of the molten pig iron inside the blast furnace.

Slag Hole—An opening in the blast furnace for the removal of the slag. Also called the cinder notch.

Slope—An inclined passage into a mine.

Sow—A main channel from which molten metal flows to the rows of pig molds on its sides. See **Pig Iron.**

Stack—A blast furnace.

Stamping Mill—Usually a building containing apparatus for crushing rock or ore. However, Hasenclever's was used to separate the particles of the iron from the cinder or slag.

Stope—An excavation from which the ore has been extracted, either above or below a level, in a series of steps. Each horizontal working is called a stope, probably a corruption of the word "step," for when several are worked, they resemble a stairway.

Strike—The direction of the outcrop of an inclined ore bed as it intersects the horizontal surface.

Surface Mining—Mining on, or near, the surface. Open-pit mining.

Tap Hole—The opening through which the molten metal is tapped or drawn from a furnace. Also called the tapping hole.

Tie Rod—A round or square rod passing through or over a furnace and connected with buckstaves to reinforce the furnace.

Tilt Hammer—A large hammer for shingling or forging iron, arranged as a lever of the first or third order and tilted or tripped by means of a cam or cog-gearing and allowed to fall upon the bar, bloom, or billet.

Timbering—Use of wooden props, posts, bars, collars, etc., to support mine workings, such as the roof and walls of a shaft or tunnel.

Trip-Hammer—A massive tilt hammer in which the lever is raised by wipers or cams. Used especially for hammering the impurities out of the iron.

Trundle Head—The open top of the blast furnace.

Tuyère—Also spelled tweet, twyer, and twere. A pipe inserted in the wall of a blast furnace through which the blast is forced into the inner chamber. Usually the tuyère enters through an embrasure in the masonry (tuyère arch).

Tymph Stone—The stone located just above the large clay plug in the front jackets of a blast furnace. It is through the clay-filled area that the tapping hole is made.

Vein—A body of ore having a more-or-less regular development in length, width, and depth.

Wrought Iron—The purest form of iron containing only about 0.05 per cent carbon; made directly from the ore, as in a bloomery or by purifying (puddling) cast iron in a reverberatory furnace or refinery.

BIBLIOGRAPHY

Allegheny Ludlum Steel. "Reunion in America," *Steel Horizons*, Vol. 1, No. 3, 1939. A quarterly publication. Article about the Pompton ironworks.

Allinson, Samuel (comp.). *Acts of the General Assembly of the Province of New Jersey from the Surrender of the Government to Queen Anne, on the 17th day of April, in the Year of Our Lord 1702, to the 14th day of January 1776*. Burlington, N.J., 1776.

American Scenic and Historic Preservation Society. *Twenty-ninth Annual Report of the American Scenic and Historic Preservation Society, 1924*. Albany, 1924. Contains an excellent illustrated article, "Historic Iron Works of the Hudson."

Appalachian Mountain Club. New York Chapter. *In the Hudson Highlands*. New York: E. G. Goldthwaite, 1945. Written by and for hikers, with a few chapters on some of the old ironworks.

Bannerman, Francis. *History of the Great Iron Chain*. 1906. Booklet.

Barrett, Walter. *The Old Merchants of New York City*. 5 Vols. New York: John W. Lovell Company, 1863–69.

Bayley, William S. "Iron Mines and Mining in New Jersey," *Geological Survey of New Jersey*, Vol. 7. Trenton, 1910. For the most part a summary of information that appeared previously in the *Annual Geological Reports*.

Beatson, Robert. *Naval and Military Memoirs of Great Britain from 1727 to the Present Time*. 6 Vols. London: J. Strachan, 1790–1804.

Beck, Rev. Henry C. In the *Newark Star-Ledger*, Dec. 15, 1957. Also see entry under "Davenport."

Beck, Lewis C. *Mineralogy of New-York*. Albany, 1842. Good local data.

Beers, Frederick W. *Atlas of Morris County, New Jersey*. New York: Beers, Ellis and Soule, 1868.

——— (ed.). *County Atlas of Orange, New York*. Chicago: A. Baskin and Burr, 1875.

Belcher, William H., and Joseph W. Belcher. *The Belcher Family*. Detroit, 1941. Very good on the Ramapo area.

Bishop, J. Leander. *A History of American Manufactures*. 2 Vols. Philadelphia: Edward Young and Company, 1864.

Boyer, Charles S. *Early Forges and Furnaces in New Jersey*. Philadelphia: University of Pennsylvania Press, 1931. Boyer was the pioneer in compiling a volume on the ironworks of New Jersey. Unfortunately, he just skimmed the surface on the history of the forges and furnaces in the northern portion of the State.

Boynton, Edward C. (comp.). *General Orders of George Washington Issued at Newburgh on the Hudson, 1782–83*. Newburgh, N.Y., 1883.

Buffet, E. P. "Hunting Old Furnaces in the Ramapo Mountains," *American Machinist*, Feb. 25, 1904, pp. 240–41; March 27, 1904, pp. 354–56; April 7, 1904, pp. 451–53. Very interesting reading.

Calendar of Historical Manuscripts Relating to the War of the Revolution. 2 Vols. New York: State of New York, 1868.

Christie, William Wallace. "Some Old-Time Water Wheels," *Engineering News*, Dec. 21, 1899. With a supplement of excellent photographs of some old forges and mills by Vernon Royle.

Clayton, W., and William Nelson. *History of Bergen and Passaic Counties, New Jersey*. Philadelphia: Everts and Peck, 1882.

Clinton, George. *Public Papers of George Clinton*. 10 Vols. New York and Albany: State of New York, 1899–1914. Some excellent source material on the ironworks that most writers on the subject completely overlooked.

Cole, David (ed.). *History of Rockland County, New York*. New York: J. B. Beers and Company, 1884. A fine county history with an excellent section on Ramapo by the Rev. Eben B. Cobb.

Colony, R. J. *The Magnetite Iron Deposits of Southeastern New York*. New York State Museum Bulletin 249–50. Albany, 1923. A "must" on mines in the Hudson Highlands and the Ramapos.

Cook, George H. *Geology of New Jersey*. Newark, N.J., 1868. A separate portfolio of maps was issued to accompany this book.

Cottrell, Alden T. *The Story of Ringwood Manor*. Trenton: Department of Conservation and Development, 1944. A good booklet.

Coxe, MacGrane. *The Sterling Furnace and the West Point Chain*. New York: privately printed, 1906. Originally given as a speech, this is a most valuable account for Coxe married into the Townsend family and possessed all of their old Sterling ironworks ledgers and papers, as well as Charles Clinton's original "Marble Book" of old surveys in the area.

Crayon, J. Percy. "Clinton Falls and Furnace," *The Evergreen News*. Reprinted in *The Highlander*, Oct. 1959.

Crèvecoeur, J. Hector St. John. *Voyage dans la Haute Pennsylvanie et dans l'Etat New York*. 3 Vols. Paris, 1801. Gives an interesting description of visits to some of the Ramapo ironworks. However, he must have relied on imagination and a rather faulty memory, for some of his statements are conflicting. An English translation of part of the set appeared as *An Eighteenth Century Journey Through Orange County*. Middletown, N.Y., 1937.

Davenport, Percy C. Letters quoted by Henry C. Beck, *Newark Star-Ledger*, July 8, 1951; Oct. 29, 1961.

Davis, F. A. *Atlas of Rockland County, New York*. Philadelphia: F. A. Davis and Company, 1876.

Davis, Mathew L. (ed.). *Memoirs of Aaron Burr*. 2 Vols. New York: Harper and Brothers, 1836.

Directory to the Iron and Steel Works of the United States. Philadelphia: The American Iron and Steel Institute, 1878, 1888.

Disturnell, J. *A Gazetteer of the State of New-York*. Albany, 1842.

Drew, Dick. "The Vanishing Village of Wawayanda," *The Sunday New Jersey Herald*, Newton, N.J., Nov. 11, 1962.

Eager, Samuel W. *An Outline History of Orange County*. Newburgh, N.Y.: S. T. Callahan, 1846–47. A very good early county history.

Ennis, W. D. In *The New York Evening Post,* Nov. 17, 1922.

Fackenthal, Benjamin Franklin, Jr. *The Great Chain at West Point and Other Obstructions Placed in the Hudson River During the War of the Revolution.* Allentown, Pa., 1937.

Fay, Albert H. *A Glossary of the Mining and Mineral Industry.* United States Department of the Interior Bull. No. 95. Washintgon, D.C.: Government Printing Office, 1920.

Fitzpatrick, John C. (ed.). *The Writings of George Washington.* 39 Vols. Washington, D.C.: The United States George Washington Bicentennial Commission, 1931–44.

Force, Peter (ed.). *American Archives.* 5th Ser. 3 Vols. Washington, D.C., 1848–53.

Forester, Frank. *The Warwick Woodlands; or, Things as They Were There Twenty Years Ago.* Philadelphia: T. B. Peterson and Brothers, 1850.

Freeland, Daniel Niles. *Chronicles of Monroe in the Olden Time.* New York: DeVinne Press, 1898.

Gordon, Thomas F. *Gazetteer of the State of New Jersey.* Trenton: Daniel Fenton, 1834.

———. *Gazetteer of the State of New York.* Philadelphia, 1836.

Green, Frank B. *The History of Rockland County.* New York: A. S. Barnes and Company, 1886.

Haines, Alanson A. *Hardyston Memorial—A History of the Township and the North Presbyterian Church, Hardyston, Sussex County, New Jersey.* Newton, N.J.: New Jersey Herald Print, 1888. Good section on iron.

Harte, Charles Rufus. "The River Obstructions of the Revolutionary War," *Papers and Transactions of the Connecticut Society of Civil Engineers.* 1946. Very good piece of research on the chains.

Harvey, Cornelius B. *Genealogical History of Hudson and Bergen Counties, New Jersey.* New York: The New Jersey Genealogical Publishing Company, 1900.

Hasenclever, Peter. *The Remarkable Case of Peter Hasenclever, Merchant.* London, 1773. A rare pamphlet published in his own defense, full of interesting data on his ironworks.

Headley, Russel. *The History of Orange County, New York.* Middletown, N.Y.: Van Deusen and Elms, 1908.

Hempstead, Elliot. In the *Bergen Evening Record,* Hackensack, N.J., Sept. 19, 1953.

Heusser, Albert H. *The Forgotten General.* Paterson, N.J.: The Benjamin Franklin Press, 1928. Out of print for many years. Now republished by the Rutgers University Press with notes and emendations by Hubert Schmidt as *Robert Erskine, George Washington's Map Maker.*

Hewitt, Edward Ringwood. *Ringwood Manor, the Home of the Hewitts.* Trenton, 1946. A very interesting booklet.

———. *Those Were the Days.* New York: Duell, Sloan and Pearce, 1943. On Ringwood.

Homes, Henry A. *A Notice of Peter Hasenclever, an Iron Manufacturer of 1764–69.* Albany: Joel Munsell, 1875. Pamphlet.

Horner, W. M., Jr. *Obstructions of the Hudson River During the Revolution.* Metuchen, N.J.: Charles F. Heartman, 1927. Also appeared in *American Collector,* Sept. 1926. Contains a copy of the Sterling ironworks contract for the West Point chain.

Horton, William. "Report of W. Horton to W. W. Mather on the Geology of

Orange County," *New York Geological Survey Annual Report 1839 for 1838*. 1839. A most valuable report on the mines.

Hotz, Preston E. *Magnetite Deposits of the Sterling Lake, N.Y.-Ringwood, N.J. Area*. Geological Survey Bull. 982-F. Washington, D.C.: Department of the Interior, 1952. On the technical side but very good.

Howell, William Thompson. *The Hudson Highlands*. 2 Vols. New York: privately printed for Frederic Delano Weekes, 1933–34. Howell spent a great deal of time hiking in the Hudson Highlands and gleaned much interesting data from the natives in the early 1900's.

Hyde, E. B. *Atlas of Passaic County, New Jersey*. New York: E. B. Hyde and Company, 1878. A desirable atlas, but scarce.

Johnson, Sir William. *The Papers of Sir William Johnson*. 13 Vols. Albany: The University of the State of New York, 1921–62.

Journals of the Continental Congress, 1774–89. 34 Vols. Washington, D.C., 1904–37.

Keller, Allan. "Private Enterprise Forges Sterling Forest Paradise," *New York World-Telegram and The Sun*, April 13, 1965.

Kreutzberg, E. C. "Orange County Iron Making," *Iron Trade Review*, July 17, 1924; July 31, 1924. An excellent research job.

Kuhler, Otto Augustus. *Peter Hasenclever*. Nyack, N.Y., 1944. The author of the booklet came from Hasenclever's part of Germany.

Lathrop, J. M. *Atlas of Orange County, New York*. Philadelphia: A. H. Mueller and Company, 1903.

Leake, Isaac Q. *Memoir of the Life and Times of General John Lamb*. Albany: Joel Munsell, 1857.

Lesley, J. P. *The Iron Manufacturer's Guide to the Furnaces, Forges and Rolling Mills of the United States*. New York: John Wiley, 1859. A most valuable reference tool.

Lossing, Benson J. *The Pictorial Field-Book of the Revolution*. 2 Vols. New York: Harper and Brothers, 1851–52. Unfortunately Lossing's account of the drawing-up of the contract for the West Point chain is inaccurate.

A Map Showing the Location of the Sterling Iron Estate Situate in Orange County, State of New York, With a Few Concise Remarks Upon the Mineral Resources and Advantages Possessed Over All Other Mineral Regions, As Regards Cheapness of Production, Facilities to Market, and Superior Quality of Metal, Submitted to the Consideration of Capitalists Engaged in the Manufacture of Iron. New York, 1856. A descriptive booklet published by the Townsends in an attempt to sell their ironworks. In the American Iron and Steel Institute.

Martin, Joseph Plumb. *Private Yankee Doodle*, ed. George F. Scheer. Boston: Little, Brown and Company, 1962.

Mather, William W. *Geology of New-York, Part I, Comprising the Geology of the First Geological District*. Albany, 1843. Good on the mines.

Mining Magazine, Vols. 1–3, 1853–54. Contains numerous articles on iron manufacture.

Monks, Minnie May. *Winbeam*. New York, 1930. On the Wanaque-West Milford area.

———. In the *Paterson Morning Call*, Paterson, N.J., April 27, 1938.

Munroe, Kirk. "A Canoe Camp 'Mid Hudson Highlands," *Outing and Wheelman*, Dec. 1884. Describes and illustrates Queensboro furnace.

Munsell, W. W. *History of Morris County, New Jersey*. New York: W. W. Munsell and Company, 1882. Excellent on the ironworks.

Nelson, William. *Historical Sketch of the County of Passaic, New Jersey*. Paterson, 1877.

———. *History of the City of Paterson and the County of Passaic, New Jersey*. Paterson, N.J.: The Press Printing and Publishing Company, 1901.

———. *Indians of New Jersey*. Paterson, N.J., 1894. Extremely useful, but copies are hard to find.

———. *Passaic County Roads*. Printed, but not published, 1877.

Nevins, Allan. *Abram S. Hewitt*. New York: Harper and Brothers, 1935. Good on Mr. Hewitt and the Ringwood ironworks.

New Jersey. *Annual Reports of the Geological Survey of the State of New Jersey*. Trenton, 1855–1905.

New Jersey. *New Jersey Archives*. 1st Ser. 42 Vols. 1880–1949; 2nd Ser. 5 Vols. 1901–17.

New Jersey Historical Society. *Proceedings of the New Jersey Historical Society*. 1845 to date. Excellent material.

New York Historical Society. *Arts and Crafts in New York, 1726–1799*. 2 Vols. New York, 1938, 1954. Lists many advertisements of ironworks. A time saver.

New York State. *Engineers' Reports*. 1867–89. Contains data on railroads serving some of the ironworks and owned by them.

New York State Historical Association. *Quarterly Journals*. See Irene D. Neu, "The Iron Plantations of Colonial New York," Jan. 1952.

North Jersey District Water Supply Commission of the State of New Jersey. *Report for the Years July 1, 1926 to June 30, 1929*. On the leveling of the Freedom furnace and construction of the Wanaque reservoir.

Parrott, Richard D. A. *Cold Blast Charcoal Pig Iron Made at Greenwood, Orange County, New York, During the Civil War Period, 1861–65*. 1921. A very informative pamphlet.

Pearse, John B. *A Concise History of Iron Manufacture of the American Colonies*. Philadelphia: Allen, Lane and Scott, 1876.

Pierson, Edward F. *The Ramapo Pass*. ed. H. Pierson Mapes. Mimeographed, 1955. Very good on the Ramapo ironworks.

Place, Frank. In *The New York Evening Post*, Nov. 17, 1922.

Pumpelly, Raphael. In *Report of the Mining Industries of the United States*. Washington, D.C.: Government Printing Office, 1886. The section on the mining industry of New York and New Jersey, by Bayard T. Putnam, is very good.

Report of the Commissioners Appointed by the Legislature of the State of New-Jersey for the Purposes of Exploring the Route of A Canal to Unite the River Delaware, Near Easton, With the Passaic, Near Newark. Morristown, N.J.: Jacob Mann, 1823. Has a fine map for locating ironworks.

Ries, Heinrich. *Report on the Geology of Orange County, New York*. New York, 1895.

Rogers, Henry D. *Description of the Geology of the State of New Jersey, Being a Final Report*. Philadelphia, 1840. Good on the mines.

Roome, William. *Early Days and Early Surveys of East Jersey*. Butler, N.J., 1897.

Rosa, C. Boehm. In the *Middletown Times Herald*, Middletown, N.Y., April, May, 1938. Excellent data on Orange County mining.

Ruttenber, E. M. *Obstructions to the Navigation of Hudson's River*. Albany: Joel Munsell, 1860. A "must" for research on the chains.

Ryerson, Albert W. *The Ryerson Genealogy*. Chicago: privately printed for Edward L. Ryerson, 1916.

Ryerson, Louis Johnes. *The Genealogy of the Ryerson Family in America, 1646–1902*. New York: Press of Jenkins and McCowan, 1902.

Scovern, Peter E. "Unspoiled Wilderness," *The Sunday New Jersey Herald*, Newton, N.J., June 20, 1963.

Simms, Jeptha R. *The Frontiersmen of New York*. 2 Vols. Albany: George C. Riggs, 1882–83. A definite "must": volume I has 50 pages of valuable information on the obstructions in the Hudson River during the Revolution.

Smith, Elizabeth Oakes. *The Salamander*. New York: G. P. Putnam, 1848. A fine frontispiece of the Augusta works in ruins. The introduction is historically interesting, but the story itself much too fanciful although it pertains to ironworks in the Ramapo Valley.

Smock, J. C. *First Report on the Iron Mines and Iron Ore Districts in the State of New York*. New York State Museum Bull. No. 7. Albany, 1889.

Spafford, Horatio. *A Gazetteer of the State of New York*. Albany: H. C. Southwick, 1813.

———. *A Gazetteer of the State of New York*. Albany: B. D. Packard, 1824.

Swank, James M. (comp.). *Classified List of Railmills and Blast Furnaces in the United States*. Philadelphia, 1873.

———. *History of the Manufacture of Iron in All Ages*. 2nd ed. Philadelphia: The American Iron and Steel Association, 1892. This revised and enlarged edition has corrected most of the errors of the original one. Like Bishop, Swank covers the whole country in his writings. Besides lacking the personal touch, writers who cover too wide an area must rely heavily on others for information. One can see in his correspondence with iron manufacturers the problems that confronted Swank: some said outright they did not know the history of their ironworks and others simply guessed. Subsequent researchers who did not go to primary sources perpetuated errors.

Tompkins, Arthur S. (ed.). *Historical Record to the Close of the Nineteenth Century of Rockland County, New York*. Nyack, N.Y.: Van Deusen and Joyce, 1902.

Torrey, Raymond H. *Peter Hasenclever, A Pre-Revolutionary Iron-Master*. A pamphlet reprint from *New York History*, Vol. 17, No. 3, July 1936.

———, Frank Place, Jr., and Robert L. Dickinson. *The New York Walk Book*. 3rd ed. New York: The American Geographical Society, 1951. Most helpful in locating the trails leading to the ironworks.

Tuttle, Joseph F. "The Early History of Morris County, New Jersey," *Proceedings of the New Jersey Historical Society*, 2nd Ser., Vol. 2, No. 1, 1870.

Van Doren, Carl. *Secret History of the American Revolution*. New York: Viking Press, 1941.

Vreeland, Ethel. *Pompton Area History*. 1960. A booklet.

Walker, A. H. *Atlas of Bergen County, New Jersey*. Reading, Pa.: C. C. Pease, 1876.

Winfield, Charles H. *History of the County of Hudson, New Jersey*. New York, 1874.

Young, C. B. F. "The Great Iron Chain." *The Iron Age*, June 3, 1937.

MANUSCRIPTS

Bauman, Sebastian. Papers. In the New York Historical Society.

Beck, Lewis C. Diaries, 1836–38. Owned by John Fleming, a New York, N.Y., rare book dealer.

Budke Volume, BC 29. Manuscript Room, New York Public Library.

Carpenter, George E., in collaboration with Miriam V. Studley. "The Cannon Ball Road." Unpublished manuscript in the Newark Public Library.

Clinton, Charles. "Marble Book," 1735–49. At the County Clerk's office, Goshen, N.Y.

Erskine, Ebenezer. Papers. In the New Jersey Historical Society.

———. Diary, 1778–79. Copy at Ringwood Manor; original in England in private hands.

Erskine, Robert. Record Book, 1774–78. In the New Jersey Historical Society.

Falck, Alyce-Louise. "Captain Solomon Townsend and His Times (1746–1811)." Unpublished thesis, 1962. Fairleigh Dickinson library, Rutherford, N.J.

Haacker, Frederick C. Papers. In the New York Historical Society.

Heath, William. Letters: Nov. 18, and Dec. 2, 1776. Author's collection.

Hewitt, Abram S. Papers. In the New York Historical Society. These are somewhat faded and difficult to read. Typed copies at Cooper Union Institute are more legible.

Knox, Henry. Papers. In the Massachusetts Historical Society.

Long Pond Ironworks. Account Books, 1866–69. Owned by Lewis West of Midvale, N.J.

New Jersey. Act, 1782 (pertaining to Ringwood ironworks). In the author's collection.

Noble and Townsend. Account Book, 1768–75. In the American Iron and Steel Institute.

Rigby, W. A. Address given at Palisades Interstate Park Conference, May 3–5, 1940, Lake Sebago, N.Y.

Ringwood Ironworks. Account Book, 1760–64. In the New Jersey Historical Society. The ledger belonged to the Ogdens.

Ryerson, Abram, Jr. Map of General William Colfax's farm area at Pompton, June 1, 1822. In the Museum Room at the Pompton Public Library.

Ryerson, Martin. Inventory, 1840.

———. Will, May 24, 1833; proved March 14, 1842. Both manuscripts in the author's collection are copies.

Southfield Furnace. Account Books, 1819–87. In the author's collection.
Sterling Ironworks. Account Books, 1792–1816. In the author's collection.
Storms, John C. "The Cannon Ball Road." Paper in the Newark Public Library.
Suffern, Andrew. Inventory, 1827. In the author's collection.
Suffern, Andrew and John. Account Book, 1822–29. In the author's collection.
———. Article of agreement to erect and manage ironworks, 1808. In the author's collection.
Townsend, Solomon. Diaries, 1798–1811. In the New York Historical Society.

INDEX